Fleet Telematics

OPERATIONS RESEARCH/COMPUTER SCIENCE INTERFACES

Greenberg /*A Computer-Assisted Analysis System for Mathematical Programming Models and Solutions: A User's Guide for ANALYZE*
Greenberg / *Modeling by Object-Driven Linear Elemental Relations: A Users Guide for MODLER*
Brown & Scherer / *Intelligent Scheduling Systems*
Nash & Sofer / *The Impact of Emerging Technologies on Computer Science & Operations Research*
Barth / *Logic-Based 0-1 Constraint Programming*
Jones / *Visualization and Optimization*
Barr, Helgason & Kennington / *Interfaces in Computer Science & Operations Research: Advances in Metaheuristics, Optimization, & Stochastic Modeling Technologies*
Ellacott, Mason & Anderson / *Mathematics of Neural Networks: Models, Algorithms & Applications*
Woodruff / *Advances in Computational & Stochastic Optimization, Logic Programming, and Heuristic Search*
Klein / *Scheduling of Resource-Constrained Projects*
Bierwirth / *Adaptive Search and the Management of Logistics Systems*
Laguna & González-Velarde / *Computing Tools for Modeling, Optimization and Simulation*
Stilman / *Linguistic Geometry: From Search to Construction*
Sakawa / *Genetic Algorithms and Fuzzy Multiobjective Optimization*
Ribeiro & Hansen / *Essays and Surveys in Metaheuristics*
Holsapple, Jacob & Rao / *Business Modelling: Multidisciplinary Approaches — Economics, Operational and Information Systems Perspectives*
Sleezer, Wentling & Cude/*Human Resource Development And Information Technology: Making Global Connections*
Voß & Woodruff / *Optimization Software Class Libraries*
Upadhyaya et al / *Mobile Computing: Implementing Pervasive Information and Communications Technologies*
Reeves & Rowe / *Genetic Algorithms—Principles and Perspectives: A Guide to GA Theory*
Bhargava & Ye / *Computational Modeling And Problem Solving In The Networked World: Interfaces in Computer Science & Operations Research*
Woodruff / *Network Interdiction And Stochastic Integer Programming*
Anandalingam & Raghavan / *Telecommunications Network Design And Management*
Laguna & Martí / *Scatter Search: Methodology And Implementations In C*
Gosavi/ *Simulation-Based Optimization: Parametric Optimization Techniques and Reinforcement Learning*
Koutsoukis & Mitra / *Decision Modelling And Information Systems: The Information Value Chain*
Milano / *Constraint And Integer Programming: Toward a Unified Methodology*
Wilson & Nuzzolo / *Schedule-Based Dynamic Transit Modeling: Theory and Applications*
Golden, Raghavan & Wasil / *The Next Wave in Computing, Optimization, And Decision Technologies*
Rego & Alidaee/ *Metaheuristics Optimization via Memory and Evolution: Tabu Search and Scatter Search*
Kitamura & Kuwahara / *Simulation Approaches in Transportation Analysis: Recent Advances and Challenges*
Ibaraki, Nonobe & Yagiura / *Metaheuristics: Progress as Real Problem Solvers*
Golumbic & Hartman / *Graph Theory, Combinatorics, and Algorithms: Interdisciplinary Applications*
Raghavan & Anandalingam / *Telecommunications Planning: Innovations in Pricing, Network Design and Management*
Mattfeld / *The Management of Transshipment Terminals: Decision Support for Terminal Operations in Finished Vehicle Supply Chains*
Alba & Martí/ *Metaheuristic Procedures for Training Neural Networks*
Alt, Fu & Golden/ *Perspectives in Operations Research: Papers in honor of Saul Gass' 80th Birthday*
Baker et al/ *Extending the Horizons: Adv. In Computing, Optimization, and Dec. Technologies*
Zeimpekis et al/ *Dynamic Fleet Management: Concepts, Systems, Algorithms & Case Studies*
Doerner et al/ *Metaheuristics: Progress in Complex Systems Optimization*

Asvin Goel

Fleet Telematics

Real-time management and planning of commercial vehicle operations

Springer

Asvin Goel
Zaragoza Logistics Center
Spain

Series Editors:
Ramesh Sharda
Oklahoma State University
Stillwater, Oklahoma, USA

Stefan Voß
Universität Hamburg
Germany

ISBN: 978-0-387-75104-7 e-ISBN: 978-0-387-75105-4

Library of Congress Control Number: 2007934783

Printed on acid-free paper.

9 8 7 6 5 4 3 2 1

springer.com

Dedicated to all victims of violence and injustice.

If you see racism or intolerance, you must act.
If you don't see them, you must open your eyes.

Preface

This book was motivated by my practical experience working with a German motor carrier specialised in so-called *road feeder services*, i.e. the road transport of air-cargo. During this work I realised that communication between dispatchers and drivers in small and medium-sized companies is often entirely realised by voice communication and that many decisions are often made manually with only basic support by computer-based decisions support tools. The lack of timely and reliable information about current vehicle positions and states certainly creates challenges in updating vehicle tours taking into account the dynamic nature of transportation processes as well as new transportation requests arriving with short advance notice. Although fleet telematics is widely recognised as the solution to improve the efficiency of commercial vehicle operations, it appears that the potentials of fleet telematics systems are currently not sufficiently exploited. This book, which was prepared as doctoral dissertation at the Chair of Applied Telematics and e-Business (University of Leipzig, Germany), seeks to show how fleet telematics systems can be used to support real-time monitoring, control, and planning of commercial vehicle operations.

This work has benefited from discussions and collaborations with many different people. I especially thank my supervisor Volker Gruhn for his support and the productive research environment I found at the Chair of Applied Telematics and e-Business. From my personal experience I know that a research environment like this cannot be taken for granted and I also want to thank my colleagues, whom I enjoyed working and spending time with. I want to express special thanks to Andrea Parosanu, Dörthe Peinelt, and Ute Noth for all their support during my stay in Leipzig and, of course, to my parents Erika and Balbir for all their support throughout my education and beyond.

Zaragoza, July 2007
Asvin Goel

Executive summary

Due to globalisation, liberalisation of markets, deregulation in the transport sector, and the increasing commitment to the just-in-time philosophy, competition between motor carriers and expectations on punctuality, reliability, flexibility, and transparency have increased significantly and will increase even more in the future. The rapid development of mobile communication and information technology allows the use of fleet telematics systems to cope with those challenges and to increase the efficiency of commercial vehicle operations. This work presents a telematics-enabled information system that alleviates a major obstacle for computer-based real-time decision support: the lack of timely and reliable information. A real-time decision support system is presented which achieves its strength from several specialised actors who collaboratively and concurrently modify problem data and solution, using different problem knowledge and solution techniques: dispatchers, a Messaging & Fleet Monitoring System, and a Dynamic Planning System. Several heuristic planning methods are presented which can be used to dynamically solve transportation problems incorporating a variety of real-life constraints that are not considered by the classical models found in the literature. Among those are the new regulations for drivers' working hours in the European Union which entered into force in April 2007. With the improved availability of timely and reliable information provided by the Messaging & Fleet Monitoring System, and the real-time decision support provided by the Dynamic Planning System, this work gives an important contribution to increasing the efficiency of commercial vehicle operations.

Contents

About the author

Asvin Goel is Visiting Associate Research Professor within the MIT-Zaragoza International Logistics Program. At the Zaragoza Logistics Center, Spain, his research activities focus on determining the value of visibility on shipments moving through the supply chain. He received his doctorate degree (Dr. rer. nat.) from the University of Leipzig, Germany. His research interests are in the area of Transport Telematics, Operations Research, and in particular Dynamic Vehicle Routing and Decision Support Systems.

Asvin has previously held courses and seminars on Transport Telematics and Operations Research at the Universities of Cologne and Leipzig and worked as an independent consultant and software developer for the motor carrier industry. For Georgi Transporte, a German motor carrier specialised in Road Feeder Services, he has redesigned and extended the existing decision support system in order to enable real-time fleet & order management and planning. Asvin has published at international conferences and journals such as European Journal of Operational Research (EJOR).

For further information please have a look at: www.fleettelematics.net

1

Introduction

1.1 Motivation

Today, more goods are transported world wide than ever before. Globalisation and liberalisation of markets will lead to even more trade in future. From 1970 to 2000 the inland transportation within the European Union (excluding the new member states) has almost doubled[1]. It appears that this considerable growth has been realised almost entirely by road transport which has almost tripled in the last 30 years. In other words, the proportion of road transport to total inland transport has grown from about one half in the year 1970 to about three quarters in the year 2000. According to a study by the European Commission[2], freight transport within the European Union (including the new member states) will increase by about 25% until 2010 and by almost 90% until 2030 compared to the values of 2000. It is assumed that this growth will also be realised almost entirely by road transport.

In some areas the volume of traffic today is already at a critical level and every day 7500 kilometres of European highways are blocked by traffic jams[3]. The possibility of extending the road networks is very limited due to social, ecological, and economical reasons. As a result, road pricing systems are likely to be increasingly deployed to reduce the level of congestion and to finance infrastructural development.

The deregulation in the European road transport market, in particular, the allowance of cabotage operations, increases competition and motor carriers from emerging countries more and more challenge motor carriers from developed countries by comparably lower wages.

Global competition forces manufacturing companies to improve the quality of their products and to reduce their manufacturing costs. As a result, manufacturing companies increasingly apply just-in-time practices in order to cut down inventory costs. Obviously, just-in-time practices necessitate punctual, reliable, and flexible

[1] See European Commission: Eurostat (2003)

[2] See European Commission: Directorate General for Energy And Transport (2003)

[3] See European Commission (2003a)

transportation, as with reduced inventory buffers any mismatch between supply and demand can result into significant disturbances of manufacturing processes.

To face those challenges motor carriers have to increase the quality of service and reduce costs. They have to increase punctuality, reliability, flexibility, and transparency of transportation services, and, at the same time, have to reduce empty mileage and low vehicle utilisation. First motor carriers have pioneered using telematics in order to cope with these challenges in the end of the eighties[1]. Other motor carriers have followed and according to a study by Frost & Sullivan[2] there were 75 550 European commercial vehicles equipped with telematics devices in the year 2001. This number is expected to rise to over 5.4 million in 2009. Rather than being a competitive advantage, the use of telematics to improve real-time management and planning of commercial vehicle operations will more and more become a necessity in order to survive in the highly competitive road transport market.

1.2 Purpose of this work

Typical commercial off-the-shelf fleet telematics systems can be used for information exchange between drivers and dispatchers, route guidance, and visualisation of vehicle positions on digital maps. They can give important information about the actual state of the transportation system which is essential for real-time management and planning of commercial vehicle operations. Many management information systems currently used by motor carriers, however, do not provide methods for processing information obtained from fleet telematics systems as, a couple of years ago, only very few commercial vehicles were equipped with telematics devices. Therefore, fleet telematics systems often cannot be easily integrated into the carrier's information system and their deployment is of only limited benefit.

This work identifies and classifies potentials of fleet telematics and shows how commercial off-the-shelf fleet telematics systems can be integrated into a typical legacy information system without telematics functionality. A Messaging & Fleet Monitoring System is presented which supports the communication between drivers and dispatchers, monitors transportation processes, determines actual data, compares actual data with planned data, and revises planned data in order to consider the actual state of the transportation system. The telematics-enabled information system alleviates a major obstacle for computer-based real-time decision support: the lack of timely and reliable information. This work presents a Dynamic Planning System (DPS) for real-time decision support which exploits the improved knowledge about the actual state of the transportation system. The real-time decision support system achieves its strength from several specialised actors who collaboratively and concurrently modify problem data and solution, using different problem knowledge and solution techniques: dispatchers, Messaging & Fleet Monitoring System, and Dynamic Planning System.

[1] See Cohen (1995)

[2] See Frost & Sullivan (2002)

The Dynamic Planning System uses algorithms to find high quality solutions to an analytical model. This model must map the real-life problem as precisely as possible as there is usually only little time to manually resolve infeasibilities resulting from an inappropriate model representation. Classical models for routing a fleet of commercial vehicles, however, oversimplify the problems that occur in practice, as pointed out by Bodin (1990) more than fifteen years ago. Although real-life problems are receiving increasing attention, this is still valid today, as stated by Kilby et al. (2000) *"More effort has gone into methods for reducing the cost of solutions than supporting rich models. However, the problems faced in industry often require rich models ..."*. This work introduces a unifying model, the General Vehicle Routing Problem (GVRP), which is a generalisation of various classical models. The GVRP is capable of considering a variety of real-life requirements such as load acceptance and employment of external carriers, time window restrictions, multiple pickup and/or delivery locations, multi-dimensional resource requirements, and a heterogeneous vehicle fleet. Although regulations regarding drivers' working hours often have a big impact on total travel times, i.e. the time required for driving, breaks, and rest periods, they have only received very little attention in the vehicle routing literature. This work shows how regulations for drivers' working hours in the European Union can be considered in vehicle routing and introduces the General Vehicle Routing Problem with Drivers' Working Hours (GVRP-DWH).

If all relevant data are known, schedules can be generated statically. In most real-life applications, however, relevant data change dynamically while vehicles are en-route. Static vehicle routing problems have been intensively studied in the vehicle routing literature. Dynamic vehicle routing problems, however, only recently have found increasing attention. This work presents two insertion methods, a Reduced Variable Neighbourhood Search algorithm, and several variants of Large Neighbourhood Search algorithms for the dynamic GVRP and GVRP-DWH. These algorithms are characterised by very fast response times and can be used within the Dynamic Planning System. In order to evaluate the proposed algorithms benchmark problems are created that incorporate many characteristics found in dynamic real-life problems. Computational experiments are performed on these benchmark problems.

With the improved availability of timely and reliable information provided by automatically analysing messages sent by vehicles, and the real-time decision support based on algorithms for solving the dynamic GVRP and GVRP-DWH, this work gives an important contribution to increasing the efficiency of commercial vehicle operations.

This work should be of particular interest to transportation professionals who want to understand how fleet telematics can be used in order to increase the efficiency of commercial vehicle operations, to developers of logistics and optimisation software who want to incorporate real-time information into their software, to producers and vendors of fleet telematics systems who want a better understanding of the requirements of their customers, and to researchers and students interested in transport telematics and operations research.

1.3 Overview

This work is organised as follows. Chapters 2 and 3 give an introduction into the general topic of this work. Chapter 4 investigates how real-time information provided by fleet telematics systems can be incorporated into management information systems used by motor carriers. Chapters 5 and 6 introduce models and optimisation methods which can be used for real-life vehicle routing problems in which data may change dynamically.

Telematics

Telematics concerns the transmission of information over a telecommunication network combined with the computerised processing of this information. Chapter 2 gives an introduction into telematics and its main enabling technologies concerned with road freight transport. Wireless communication techniques which can be used for information exchange between dispatchers and drivers are surveyed. Another fundamental enabling technology for many telematics applications is the determination of a vehicle's position. Chapter 2 surveys the fundamental positioning systems used for in-vehicle positioning. Geographical Information Systems for Transportation are briefly introduced, as they are particularly required to determine shortest routes, and to map a vehicle's position to the corresponding point in the road network. Eventually, chapter 2 surveys transport telematics applications which are of particular interest to motor carriers.

Commercial vehicle operations

Chapter 3 presents an overview over the development of road freight transport and its impact on commercial vehicle operations. The dramatic changes in the transport industry during the last decades are described and a brief look at the future development of road freight transport is given. The fundamentals of road freight transport are examined focusing on its main characteristics: transportation request, transportation resources, and the transportation services provided. Activities and management decisions of motor carriers can be categorised according to their impact on future operations. Chapter 3 discusses the different management levels: strategic, tactical, operational, and real-time management. Eventually, operational and real-time tasks are discussed in more detail before a case study is presented. In the following chapters the models and methods presented are also put in context to this case study.

Management information systems

Chapter 4 investigates management information systems used by motor carriers to perform their tasks at the operational and real-time management level. Many management information systems currently used do not have any telematics functionality as, a couple of years ago, only very few commercial vehicles were equipped with telematics devices. Chapter 4 briefly describes such a typical legacy information system, focusing on those functions affected by the communication possibilities between drivers and dispatchers. Functionalities provided by fleet telematics systems are described and potentials arising with the use of such systems are identified and classified. It is shown how commercial off-the-shelf fleet telematics systems can be integrated into a typical legacy information system without telematics functionality. A Messaging & Fleet Monitoring System is presented which supports the communication between drivers and dispatchers, monitors transportation processes, determines actual data, compares actual data with planned data, and revises planned data in order to consider the actual state of the transportation system. The lack of timely and reliable information used to be a major obstacle for computer-based real-time decision support. Chapter 4 presents a Dynamic Planning System which can be used to provide real-time decision support considering the improved knowledge about the actual state of the transportation system. A transaction control scheme is presented allowing dispatchers, Messaging & Fleet Monitoring System, and Dynamic Planning System to collaboratively and concurrently modify problem data and solution, using different problem knowledge and solution techniques. Directions for extending the telematics-enabled information system by additional functionalities provided by electronic freight markets are given. Chapter 4 concludes with a presentation of the implementation of the Messaging & Fleet Monitoring System and a prototype of the Dynamic Planning System.

Models for routing a fleet of commercial vehicles

Chapter 5 surveys classical models for routing a fleet of commercial vehicles and presents mathematical formulations of these models. Real-life vehicle routing problems encounter a variety of practical complexities which, to a certain extend, have been considered by the classical models. However, the classical models often oversimplify the problems that occur in practice. Chapter 5 introduces a general model, that can handle the requirements evolving from various characteristics found in real-life vehicle routing problems that are not considered by the classical models. This model, which will be termed the General Vehicle Routing Problem (GVRP), unifies the formulations of the Vehicle Routing Problem, the Pickup and Delivery Problem, and various variants and generalisations of these problems. Although regulations regarding drivers' working hours often have a big impact on total travel times, i.e. the time required for driving, breaks, and rest periods, they have only received very little attention in the vehicle routing literature. Chapter 5 shows how regulations for drivers' working hours in the European Union can be considered in vehicle routing and scheduling and introduces the General Vehicle Routing Problem with Drivers' Working Hours (GVRP-DWH). Chapter 5 concludes by showing how the problem the motor carrier of the case study has to face can be modelled as a GVRP-DWH.

Dynamic vehicle routing

The construction of schedules is a key issue for motor carriers and their success is highly dependent on the generation of good schedules. If all relevant data are known a priori, schedules can be generated statically. In most real-life applications, however, relevant data change while vehicles are en-route and schedules have to be updated dynamically. Chapter 6 investigates the main differences between dynamic and static planning. Algorithms developed for the classical models are surveyed, focusing on those that are suitable for rich vehicle routing problems in which data may change dynamically. Neighbourhood operators which allow to move from one feasible solution of the GVRP or GVRP-DWH to another feasible solution are introduced. Chapter 6 presents two insertion methods that can be used to quickly improve a solution considering new transportation requests arriving dynamically. Furthermore, a Reduced Variable Neighbourhood Search algorithm, which achieves its strength from changing the neighbourhood structure during the search, and several Large Neighbourhood Search algorithms, which iteratively remove an re-insert some of the transportation requests, are presented. The algorithms presented are characterised by very fast response times and can be used within the Dynamic Planning System. Computational experiments are performed to evaluate the algorithms presented.

Conclusions

Chapter 7 gives a summary of this work and a discussion of the scientific contributions. Eventually, some directions for future research are given.

2

Telematics

2.1 Introduction

The term *telematics* describes the combination of the transmission of information over a telecommunication network and the computerised processing of this information. It is the anglicised version of the French word *télématique* which is a merger of the words *télécommunication* and *informatique* and has been coined 1978 by Simon Nora and Alain Minc in their report titled *L'Informatisation de la société*[1]. This report was mandated by the French president Valéry Giscard d'Estaing in 1976 who was solicitous that *"the applications of the computer have developed to such an extent that the economic and social organisation of our society and our way of life may well be transformed as a result"*.

Recent developments of computer and telecommunication technology have an equally important impact on society and economy today as the increasing availability of small and affordable personal computers in the seventies. As computers are becoming much smaller and less energy-hungry, computing devices are becoming mobile and pocket computers can accompany us wherever we are. Telecommunication technology can be embedded in those mobile devices enabling wireless telecommunication with stationary devices and other mobile devices. The emergence of new fields of application has resulted in new branches of computer science sometimes described with the terms *mobile computing*, *ubiquitous computing*, or *pervasive computing*. With the recent developments in wireless communication and portable computing devices, there is a shift in the interpretation of the term telematics towards applications based on wireless communication. In addition, it is often presumed that at least one computing device is involved which is not a conventional computer or laptop. Throughout this work the focus will be on telematics applications according to this interpretation of the term telematics.

This chapter gives an introduction into the enabling technologies for telematics applications, in particular those concerned with transportation. A definition of the

[1] See Nora and Minc (1978)

term *transport telematics* is given and those transport telematics applications which are of particular concern to motor carriers are surveyed.

2.2 Enabling technologies

In this section the most important enabling technologies for telematics applications concerning commercial vehicle operations are surveyed, i.e. wireless communication, positioning systems, and Geographical Information Systems.

2.2.1 Wireless communication

Wireless communication is a prerequisite for information exchange between vehicles or drivers, and stationary systems. Wireless communication is primarily realised using electromagnetic waves, however, short distances can also be bridged using infrared communication. The coverage of very large areas encounters several problems which originate from the characteristics of electromagnetic waves. In an idealised scenario electromagnetic waves spread equally in all directions and their intensity reduces quadratically with the distance to the transmitter. In real-life, reflections, absorptions, scattering, refractions, and electromagnetic perturbations reduce the intensity of electromagnetic waves significantly. Thus, the reduction of intensity reduces in the fourth power with the distance[1]. That is, in order to double the geographic coverage, a 16 times stronger transmitter is needed. As electromagnetic waves with equal frequencies sent by different transmitters interfere with another, radio communication requires the reservation of the used radio frequencies. However, frequency ranges are limited and wireless communication techniques have to appropriately deal with this problem. In this section the most important wireless communication techniques are described.

2.2.1.1 Trunked radio

Conventional radio communication requires the reservation of radio frequencies for each user group. The licensed radio frequencies are only used by one user group and each user group must license their own frequency. To deal with the increasing demand and the finite amount of available radio spectrum, trunked radio system use several frequencies which are allocated to individual users on demand. This allows for more efficient utilisation of limited frequencies because each user group does not require a dedicated channel.

TETRA (**TE**rrestrial **T**runked **Ra**dio)[2], is a European standard for modern digital trunked radio defined by the European Telecommunications Standards Institute (ETSI). For civil systems in Europe the frequency bands 385-390 MHz, 395-399.9 MHz, 410-430 MHz, 450-470 MHz, 870-876 MHz, and 915-921 MHz, have been allocated for TETRA by ERC Decision (96)04[3]. Data transfer with TETRA is

[1] See Freeman (1987)
[2] TETRA was formerly known as abbreviation of **T**rans **E**uropean **T**runked **Ra**dio
[3] See European Radiocommunications Committee (1996)

at 7.2 kbps. Due to the low frequency used, high levels of geographic coverage can be achieved with a smaller number of transmitters. TETRA was developed to meet the needs of organisations and companies who need fast one-to-one and one-to-many voice and data communication in their daily work. Users of trunked radio communication are typically public safety and security organisations such as police, fire and rescue forces, but also other professional user groups such as commercial vehicle fleets.

2.2.1.2 Cellular communication

In cellular communication networks the covered area is partitioned into multiple cells and each cell is serviced by its own low range transmitter. Each mobile telephone communicates with a transmitter within range and the information is forwarded within the cellular network towards the recipient. Cellular communication networks have the advantage that only a small distance to the stationary transmitters has to be bridged. Furthermore, transmitters which are far apart can use the same frequencies as illustrated in figure 2.1. When a mobile user travels from one cell to another the communication link has to be reconfigured. As neighbouring cells use

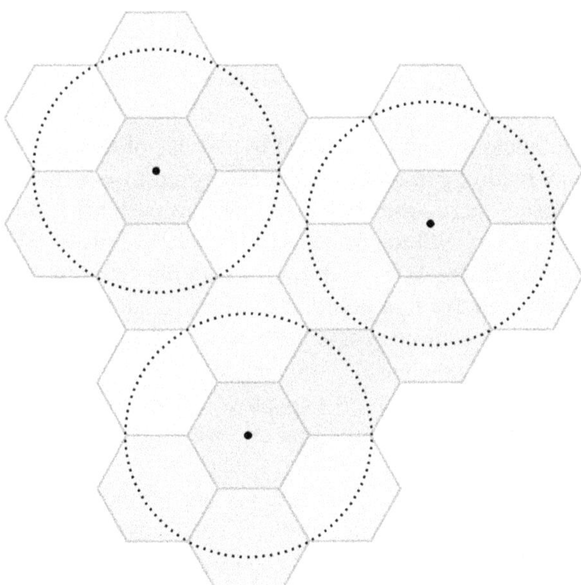

Fig. 2.1: Base stations of cells with the same colour can use the same frequencies due to the limited range of the signals

different frequencies, the mobile device switches its frequency to the new cell. This reconfiguration is called *handover*.

First generation mobile communication system were introduced in the eighties and were analog systems primarily developed for voice communication.

The *Global System for Mobile Communication* (GSM) differs significantly from its predecessors. Both signalling and speech channels are digital, which means that it is seen as a second generation mobile communication system. For GSM the frequency bands 890-915 MHz and 935-960 MHz have been allocated by ERC Decision (94)01[1]. GSM allows bitrates of 9.6 kbps for data transfer.

Second generation cellular communication networks were built mainly for telephone calls and only had slow data transmission capabilities. Due to the rapid changes in technology, these factors do not meet the requirements of today's wireless revolution. The *Universal Mobile Telecommunications System* (UMTS) is a third generation mobile communication system allowing much higher bitrates. UMTS is designed with both terrestrial and global satellite components. For terrestrial UMTS the frequency bands 1900-1980 MHz, 2010-2025 MHz and 2110-2170 MHz, and for satellite UMTS the frequency bands 1980-2010 MHz and 2170-2200 MHz have been allocated by ERC Decision (97)07[2]. The bitrate is 144 kbps for full outdoor mobility applications in all environments, 384 kbps for limited-mobility outdoor applications in the micro and macro cellular environments (in urban and suburban areas), and 2048 Mbps for low-mobility outdoor applications, particularly in the pico and micro cellular environments (in indoor and urban areas).

2.2.1.3 Satellite communication

Voice and data communication can be realised by the use of communication satellites. Satellite communication can be classified according to whether the satellites are positioned in a geostationary orbit (GEO) or low-Earth orbit (LEO).

GEO satellites orbit in an altitude of 35785 kilometres above the Earth. In this height they move at a speed which is synchronous with the circulation of the Earth. Thus, they are stationary relative to a point on the Earth's surface. Due to the large distance to the Earth each geostationary satellite can cover a huge area, see figure 2.2. Assuming a minimum ground antenna elevation angle of 10 degrees, a single satellite in geostationary orbit can cover about 34 percent of the Earth's surface. The large altitude of a geostationary satellite results in a one-way time delay of at least 0.25 seconds[3].

Geostationary satellite communication systems are provided by Inmarsat and Qualcomm[4]. Inmarsat operates nine geostationary satellites which provide global coverage. Only four of the satellites are active, and five are for emergency back-up. Communications via the Inmarsat-C system are data or message-based. Messages are

[1] See European Radiocommunications Committee (1994)

[2] See European Radiocommunications Committee (1997)

[3] See Comparetto and Ramirez (1997)

[4] See Inmarsat (2005) and Qualcomm (2005)

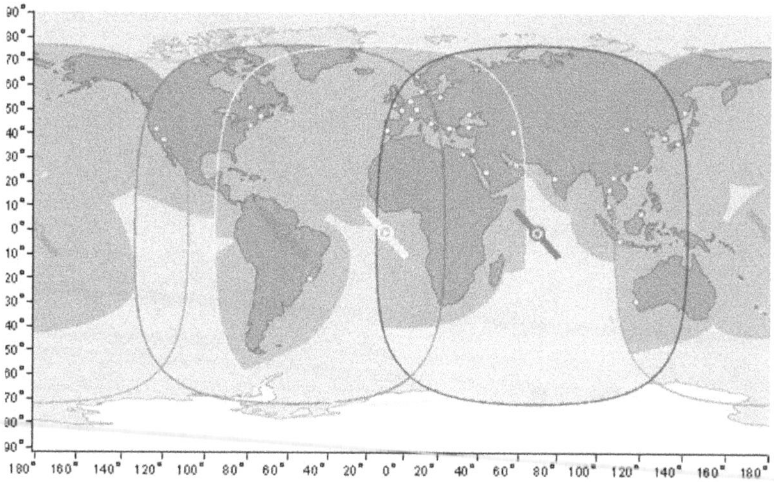

Fig. 2.2: Global coverage with four geostationary satellites

transferred to and from an Inmarsat-C terminal at an bitrate of 600 bps. Frequencies are 1530.0-1545.0 MHz (downlink) and 1626.5-1645.5 MHz (uplink). The EutelTRACS service provided by Qualcomm is realised by two satellites covering Europe, the Mediterranean, an the Middle East. The satellites operate on the frequency bands 10.70-11.70 GHz and 12.50-12.75 GHz (downlink) and 14.00-14.25 GHz (uplink) providing low bitrate data communications. The downlink bitrate is between 5 kbps and 15 kbps while the uplink is between 55 bps and 165 bps.

Low-Earth orbit satellite communication systems use satellites which are in much lower orbits than geostationary satellites. Due to the low orbits, those satellites are not geostationary and orbit the Earth in 1.5 to 10 hours depending on the height. They provide a small geographic coverage and thus, more satellites are required if continuous coverage is desired. Due to the lower distance to the Earth, less intense and smaller transmitters are required - for both the satellites and the ground side systems. ORBCOMM[1] provides LEO communication systems with 35 satellites orbiting in a height of about 775 kilometres. As figure 2.3[2] illustrates, global coverage is not continuously. Short gaps in the coverage are closed by one of the passing satellites in a few minutes, providing global coverage with latency. The satellites operate at frequencies of 137.00-138.00 MHz (downlink) and 148.00-150.05 MHz (uplink). The downlink bitrate is at 4.8 kbps while the uplink is at 2.4 kbps.

---

[1] See ORBCOMM (2005)

[2] The illustration has been rendered by SaVi - Wood et al. (1996)

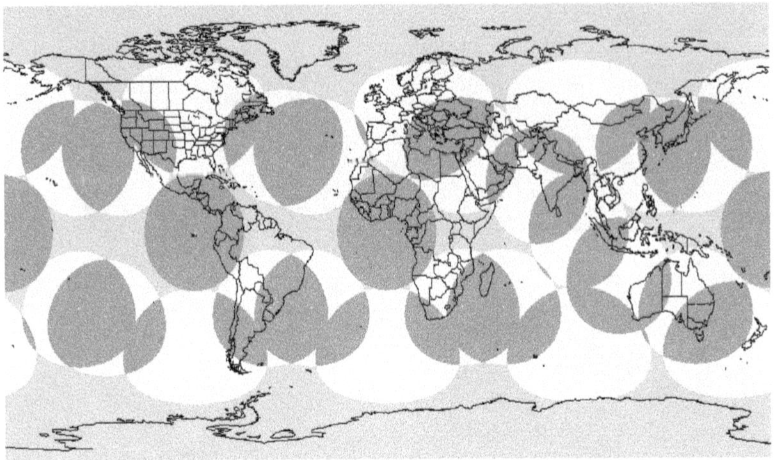

Fig. 2.3: Low-Earth satellite communication systems achieve global coverage only with latency

2.2.1.4 Dedicated Short Range Communication

Dedicated Short Range Communications (DSRC) is a short to medium range wireless communication technique specifically designed for automotive use, i.e. vehicle-to-vehicle and vehicle-to-infrastructure communication. Due to the short range of the signals, DSRC is particular useful to provide location based services. Today, the main application of DSRC is Electronic Toll Collection (ETC). In future, DSRC will also support safety critical communications such as collision avoidance and hazard warning. DSRC systems use infrared or the radio spectrum, particularly microwaves in the frequency bands 5.795-5.805 GHz and 5.805-5.815 GHz[1].

2.2.1.5 Broadcasting

Broadcasting is mainly used for the distribution of *traffic and travel information*. The *Radio Data System* (RDS) is a standard for sending small amounts of digital information using conventional FM radio broadcasts. RDS uses the technique of adding data at a bitrate of 1187.5 bps on an existing stereo transmission in a way that the data is carried inaudibly. The *Traffic Message Channel* (TMC) is a service of the RDS which provides traffic information coded according to the ALERT-C protocol[2]. TMC messages are processed by in-vehicle RDS-TMC receivers which use this information to give route guidance considering the current traffic and weather conditions.

Digital Audio Broadcasting (DAB) is a technology for broadcasting audio in digital form. DAB was developed within the Eureka 147 Project and is now standardised by the European Telecommunications Standards Institute (ETSI). DAB uses the

[1] See Electronic Communications Committee (2002)

[2] See International Organisation for Standardization (2003)

frequencies 47-68 MHz, 174-240 MHz and 1452-1492 MHz. The gross data capacity for the entire DAB signal is approximately 3 Mbps, of which approximately 2.3 Mbps can be used for data transmission. Considering redundancy in channel encoding, a net useful payload in the range of 0.6-1.7 Mbps is available[1]. As the bitrates are magnitudes higher than those available with RDS-TMC, more sophisticated traffic and travel information can be broadcasted using DAB.

2.2.2 Positioning systems

Determining the position of vehicles is a fundamental task in transportation. The knowledge of vehicle positions is important for autonomous navigation, collective traffic observation, and tracking of commercial vehicles. This section presents an overview of positioning systems which can be used in commercial vehicles.

2.2.2.1 Dead reckoning

If the vehicle's position is known at one point in time, the position can be continuously determined by advancing the known position using course, speed, time and distance travelled. This technique is known as dead reckoning.

The vehicle's speed and the distance travelled can be determined using wheel odometers. Each turn of the wheel is identified and the distance travelled can be determined by the circumference of the wheels. Odometer inaccuracies result from wear and slip of the wheels. The course can be determined using magnetic or gyroscopic compasses. As the magnetic field of the Earth is very weak, the accuracy of the course determined by magnetic compasses is subject to all kinds of magnetic perturbations. Gyroscopic compasses use mechanical or optical gyroscopes to determine the course of a vehicle. A mechanical gyroscope consists of a rapidly spinning wheel set in a framework that permits it to tilt freely in any direction or to rotate about any axis. The momentum of such a wheel causes it to retain its attitude when the framework is tilted. An optical gyroscope, laser or fibre, measures the interference pattern generated by two light beams, travelling in opposite directions within a mirrored ring or fibre loop, in order to detect very small changes in motion[2].

The advantage of dead reckoning is that it allows fully autonomous positioning within the vehicle. The main disadvantage of dead reckoning is its unbounded accumulation of errors. Thus, dead reckoning requires a method for position correction such that errors accumulated since the last correction can be eliminated. An extensive discussion of dead reckoning sensors as well as methods for position correction is out of scope of this work and can be found in Czommer (2000).

2.2.2.2 Satellite positioning

A *Global Navigation Satellite System* (GNSS) allows a mobile receiver to determine its exact position anywhere in the world. Currently, there are three GNSS among

[1] See Bower (1998)

[2] See Britannica Online (2005)

which only the first one is fully operational: the United States' *Global Positioning System* (GPS), the Russian Federation's *Global'naya Navigatsionnaya Sputnikovaya Sistema* (GLONASS), and the European Union's *Galileo*. All GNSS use trilateration to locate a mobile receiver through calculations involving information from a number of satellites.

Satellite positioning of vehicles relies on the knowledge of the exact position of satellites and the distance of the vehicle to those satellites. Let d_{vs} denote the distance of vehicle v to satellite s and let (x_s, y_s, z_s) denote the satellite's position in space. The position (x_v, y_v, z_v) of vehicle v can be calculated with the help of the following equation.

$$d_{vs} = \sqrt{(x_s - x_v)^2 + (y_s - y_v)^2 + (z_s - z_v)^2}$$

This equation has three variables and three satellites are sufficient to determine the position of the vehicle. As illustrated in figure 2.4, two spheres intersect in a circle. The intersection of three spheres results in two distinct positions. Only one of them is near the Earth's surface whereas the other is far in space and can be discarded.

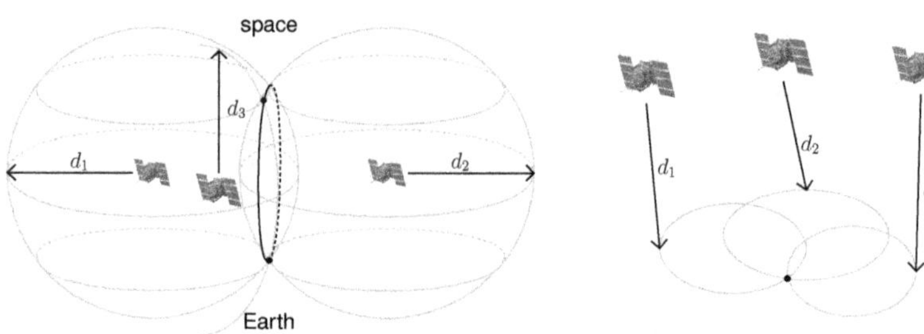

Fig. 2.4: Satellite positioning with three satellites

In an ideal scenario each satellite s sends a signal which includes the exact time t_s of transmission. The signal travels with the speed of light c (which is approximately 300 000 km/s) towards the receiver where it arrives at the time t_v. The distance between satellite and receiver is

$$d_{vs} = c \cdot (t_v - t_s).$$

In real-life, however, the clocks of satellites and receivers in the vehicles are not always running synchronously. Due to the high speed of light an error of 1 μs results in a difference of 300 metres. The clocks of the satellites are very precise and are regularly synchronised by ground control. The clocks of the receivers, however, are usually not as precise and are not synchronised with the satellite clocks. Fortunately, the precise time of the internal clocks of the receiver is not required. Instead, the time is calculated using a fourth satellite signal. Let $\tilde{t_s}$ denote the locally determined time of satellite s and let $\delta_s := \tilde{t_s} - t_s$ denote the time bias. Analogously, let $\tilde{t_v}$ denote the locally determined time of the receiver and let $\delta_v := \tilde{t_v} - t_v$ denote the time bias. The pseudo range $\tilde{d}_{vs}$ based on the locally determined time can be used to determine the position by

$$\begin{aligned}\tilde{d}_{vs} := c \cdot (\tilde{t_v} - \tilde{t_s}) &= c \cdot (t_v - t_s) + c \cdot (\delta_v - \delta_s) \\ &= d_{vs} + c \cdot (\delta_v - \delta_s) \\ &= \sqrt{(x_s - x_v)^2 + (y_s - y_v)^2 + (z_s - z_v)^2} + c \cdot \underbrace{(\delta_v - \delta_s)}_{=:w_v}.\end{aligned}$$

As the clocks of all satellites are synchronised regularly each satellite s has approximately the same time bias δ_s. If $\delta_v - \delta_s$ is substituted by w_v the above equation has four variables and four satellites are sufficient to determine the position of the vehicle.

Accuracy

Satellite positioning is subject to several influences having effect on the quality of positioning. The accuracy of satellite positioning suffers from the following influences:

- Satellite clocks
  Although atomic clocks used in satellites are very precise, no clock is absolutely precise, and the clock error continuously grows between subsequent synchronisations by ground control.
- Satellite orbits
  Satellites are positioned in very precise orbits, however, slight shifts of the orbits are possible due to gravitation forces.
- Atmospheric effects
  Satellite signals do not travel at the vacuum speed of light as they transit the ionosphere and troposphere. Free electrons in the ionosphere as well as variations in temperature, pressure, and humidity contribute to the speed of radio waves.
- Multi-path effects
  Satellite signals can be reflected by high rise buildings and other obstacles. In urban areas the probability that satellite signals cannot reach the receiver on the direct line is very high, in particular, if the satellite is in a low horizon.

Differential GNSS[1] uses the fact that inaccuracies caused by those influences can be expected to be similar for receivers located near to each other. In order to improve the accuracy of positioning of a vehicle, a second receiver located at a fixed known position can be used. The second receiver is used to measure the signal error. This allows to calculate corrections which can also be applied to the position obtained by the vehicle. Let (x_b, y_b, z_b) denote the known position of the base station and $(\tilde{x}_b, \tilde{y}_b, \tilde{z}_b)$ denote the calculated position from the satellite signals. Analogously, let

[1] Differential GNSS is often referred to as *differential GPS* as today GPS is the only fully operational GNSS.

$(\tilde{x}_v, \tilde{y}_v, \tilde{z}_v)$ denote the calculated position of the vehicle. The difference between the measured position and the exact position can be expected to be the same for the base station as for the vehicle. The vehicle's position (x_v, y_v, z_v) can then be calculated by

$$(x_v, y_v, z_v) = (\tilde{x}_v + x_b - \tilde{x}_b, \tilde{y}_v + y_b - \tilde{y}_b, \tilde{z}_v + z_b - \tilde{z}_b).$$

This correction, however, is only effective if the same satellites are used for calculating the position. Correction data is usually transmitted via one-way broadcasting and thus, it is not known which satellite combination is used by the vehicle. As the number of satellite combinations which may be used for positioning is very high, it is not practical to transmit correction data for all satellite combinations to the receiver. Instead, correction data Δr_{bs} for the pseudo ranges of all satellites is transmitted. The pseudo ranges measured by the base station b and the vehicle are

$$\begin{aligned}\tilde{d}_{bs} &:= c \cdot (\tilde{t_b} - \tilde{t_s}) \\ &= d_{bs} + \underbrace{c \cdot (\delta_b - \delta_s) + \varepsilon_{bs}}_{=:\Delta r_{bs}}\end{aligned}$$

and

$$\begin{aligned}\tilde{d}_{vs} &:= c \cdot (\tilde{t_v} - \tilde{t_s}) \\ &= d_{vs} + c \cdot (\delta_v - \delta_s) + \varepsilon_{vs}\end{aligned}$$

where ε_{bs} and ε_{vs} denote the error due to the various influences. It is assumed that for nearby receivers the errors ε_{bs} and ε_{vs} are almost identical and that the position can be calculated using the adjusted pseudo ranges by:

$$\begin{aligned}\hat{d}_{vs} := \tilde{d}_{vs} - \Delta r_{bs} &= d_{vs} + c \cdot (\delta_v - \delta_s) + \varepsilon_{vs} - \Delta r_{bs} \\ &= d_{vs} + c \cdot (\delta_v - \delta_s) + \varepsilon_{vs} - c \cdot (\delta_b - \delta_s) - \varepsilon_{bs} \\ &= d_{vs} + c \cdot (\delta_v - \delta_b) + \underbrace{\varepsilon_{vs} - \varepsilon_{bs}}_{\approx 0} \\ &\approx d_{vs} + c \cdot (\delta_v - \delta_b) \\ &= \sqrt{(x_s - x_v)^2 + (y_s - y_v)^2 + (z_s - z_v)^2} + c \cdot \underbrace{(\delta_v - \delta_b)}_{=:w_v}.\end{aligned}$$

Under the assumption that $\varepsilon_{bs} = \varepsilon_{vs}$ an equation with four variables has to be solved and four satellites are sufficient to determine the position.

Availability

Satellite positioning requires the "visibility" of the satellites. Due to tunnels, urban canyons, and other obstacles, vehicles cannot always receive the signals of four satellites. Furthermore, it is not always guaranteed that the signals received are sufficiently precise due to multi-path effects. As illustrated in figure 2.5, insufficient satellite visibility can result in poor positioning. Therefore, satellite positioning systems usually cannot be used to fully replace dead reckoning systems. Instead, they should be used in conjunction with dead reckoning systems in order to provide high availability of accurate positioning.

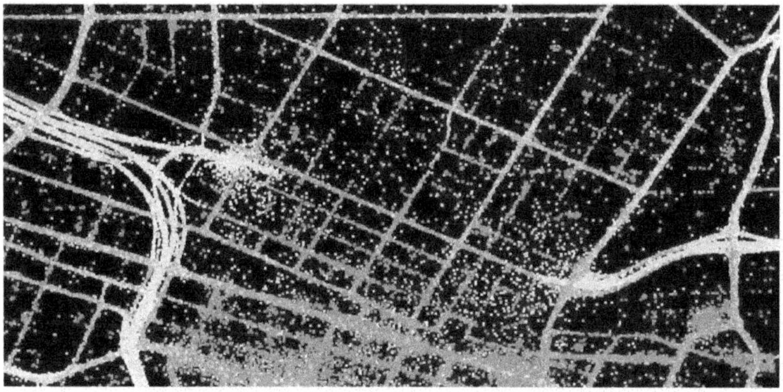

Source: Mason (2005)

Fig. 2.5: Obstacles such as tunnels hinder satellite positioning

2.2.2.3 Cellular communication based positioning

In cellular communication networks base stations are distributed throughout the covered area. There is a variety of ways in which a position can be determined in cellular communication networks such as GSM and UMTS. According to Drane et al. (1998), the most important techniques for positioning in cellular communication networks are cell of origin (COO), propagation time, time difference of arrival (TDOA), and angle of arrival (AOA) which are illustrated in figure 2.6.

Cell of origin (COO)

The simplest but also most inaccurate way is to approximate the vehicle's position by the position of the COO, i.e. the cell which is used for communication. The COO only gives an approximation of the vehicle's position and, as illustrated in figure 2.6, it is not guaranteed that the base station associated to the COO represents the closest base station to the vehicle. In the GSM network COO can give an accuracy of less than 100 metres in urban areas and up to 35 kilometres in rural areas.

Propagation time

This involves measuring the time it takes for a signal to travel between a base station and a mobile telephone or vice versa. Alternatively, this approach might involve the measurement of the round-trip time of a signal transmitted from a source to a destination which is then echoed back to the source, giving a result twice that of the one-way measurement. The former requires very stable and accurate clocks and the knowledge of the exact time of signal transmission. The latter does not rely on such synchronisation between the mobile telephone and the base station(s) and thus, is the more common means of measuring propagation time. Three base stations are required to give unambiguous positioning.

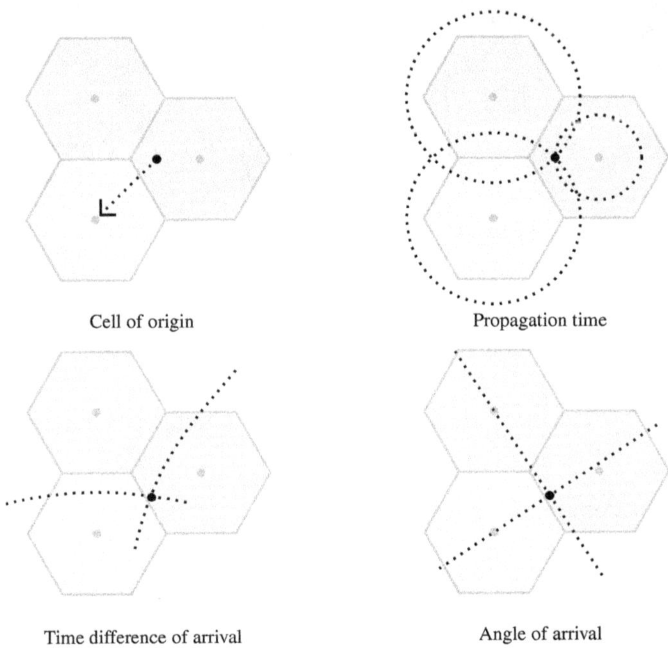

Fig. 2.6: Cellular communication based positioning

Time difference of arrival (TDOA)

A mobile telephone can listen to signals transmitted simultaneously by several base stations and measure the time difference between each pair of arrivals. Each TDOA measurement defines a curve on which the mobile telephone must be located. Let (x_A, y_A) and (x_B, y_B) denote the known positions of the base stations A and B and let Δt_{AB} denote the TDOA. The position (x_v, y_v) can be determined by

$$\sqrt{(x_A - x_v)^2 + (y_A - y_v)^2} + c \cdot \Delta t_{AB} = \sqrt{(x_B - x_v)^2 + (y_B - y_v)^2}.$$

Two or three TDOA measurements are required for unambiguous positioning. An important issue for TDOA systems is the need to have some means of establishing the synchronicity of the base stations. For self-positioning the base station must transmit the signal at the same time (or with a known time offset), for remote positioning the signal transmitted from the mobile telephone is received by several base stations and there must be a known time relationship between the receiver clocks.

Angle of arrival (AOA)

This involves measuring the AOA of a signal from a base station at a mobile telephone or the AOA of a signal from the mobile telephone at a base station. In either case a single measurement produces a straight line. If the mobile telephone is not on the direct line through two different base stations, both lines intersect at the vehicle's position.

2.2.2.4 Signpost systems

Signpost systems can be used for positioning of vehicles as they pass roadside beacons. Vehicles and beacons are equipped with DSRC devices and when a vehicle passes a signpost, it receives encoded locational identifier information from the signpost. Positioning using signpost systems relies on a sufficient number of signposts located along the roads. As deploying the required infrastructure for widespread areas is very expensive, signposts are primarily used if other positioning methods cannot be used or if the accuracy is insufficient, e.g. in covered areas and roadways.

2.2.3 Geographical Information Systems

Geographical Information Systems (GIS) are systems for capturing, storing, checking, manipulating, analysing, and displaying data which are spatially referenced to the Earth[1]. Among the most important applications of GIS are *Geographical Information Systems for Transportation* (GIS-T)[2]. This section gives a brief introduction into GIS-T and its applications.

2.2.3.1 Data collection

Transport related data can be captured using airborne or land-based methods. Although satellites and aeroplanes can be used to obtain aerial images of the Earth's surface at relatively low cost, road mapping information required for GIS-T databases can often only be obtained by land-based methods.

Aerial images

It is possible to identify some transport related data, e.g the road network, using aerial images[3]. This of course, is only possible if the geographical features are visible from the sky. Hence, aerial images do not lend themselves well to mapping roads in dense urban canyons and tree canopy areas. Furthermore, prevailing traffic regulations cannot be identified using aerial images as road signs can not be captured.

Mobile mapping

Mobile mapping involves the use of a vehicle equipped with sensors and cameras to capture transport related data while driving. This ensures fast and low-cost data acquisition. In principle, all information that a participant in traffic is able to see can be acquired[4]. The required processing of raw data, e.g. for road sign recognition, can be done in real-time or after the collection of the raw data.

[1] See Department of the Environment (1987)
[2] See Miller and Shaw (2001)
[3] See Baumann (2002)
[4] See Benning and Aussems (1998)

2.2.3.2 Data representation

Two fundamental geographical data models, the raster model and the vector model[1], are used for representing geographical data.

Raster model

In a raster representation, the Earth's surface is divided into an array of cells that are usually square or rectangular. All persistent geographical variation is expressed by assigning attributes to these cells. These attributes can represent the type of the cell, e.g. building or road. When information is represented in raster form all details about variations within the cells are lost and the cell's attributes can only represent a simplification. For a precise representation the cell size has to be small enough in order to minimise the amount of information lost. However, a small cell size dramatically increases the storage memory required for the raster representation. Data encoded using the raster data model are particularly useful as a backdrop map display because they look like conventional maps and can communicate a lot of information quickly to humans. However, the raster representation is not very useful for computerised analysis of the road network.

Vector model

In the vector model, each object in the real world is classified into a geometric type: point, line, or area. Points are recoded by their coordinates, lines as points defining the vertices of the line, and areas as a series of lines that close to form a polygon. The vector representation of the road network is particularly useful as it is very storage efficient and well suited for various applications. Most transport related applications, e.g. route calculations, are based on a vector data model of the road network, e.g. the *Geographic Data File*.

2.2.3.3 The Geographic Data File

The *Geographic Data File* (GDF) is a standard used to describe and exchange road network-related data. Major map vendors such as TeleAtlas and NAVTEQ provide maps in GDF. The current GDF version 4.0 was published as an ISO standard in 2004[2]. The conceptual data model of GDF identifies *features, attributes* and *relationships* which are defined in catalogues[3]. The *Feature Catalogue* provides a definition of real world objects such as roads and buildings. The *Attribute Catalogue* defines a number of characteristics of features and possibility of relationships. The *Relationship Catalogue* describes relations between features that may be used, e.g. to indicate the right-of-way.

The GDF structure is organised in three levels which are illustrated in figure 2.7. In level 0 the fundamental geometrical and topological entities used are described.

[1] See Longley et al. (2001)
[2] See International Organisation for Standardization (2004)
[3] See Comité Européen de Normalisation: Technical Committee 278 (1995)

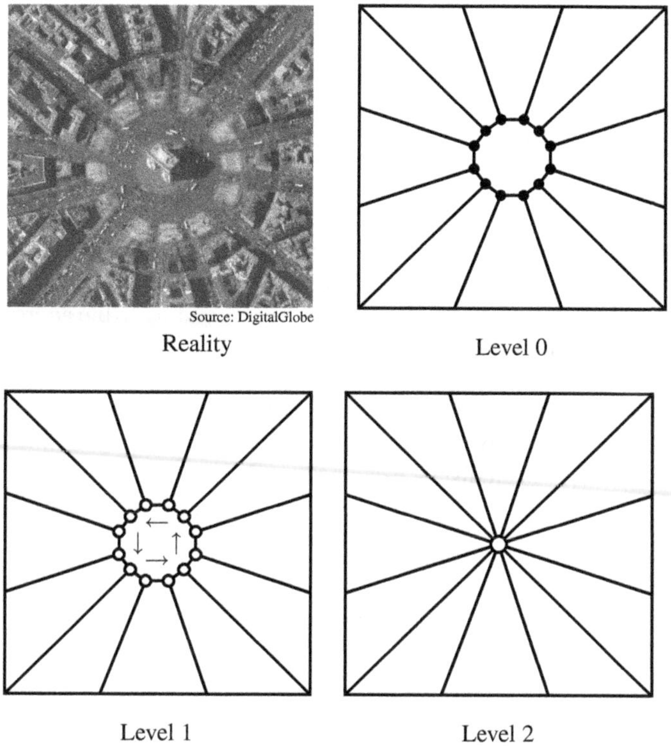

Fig. 2.7: GDF levels

The entities are nodes (0-dimensional), edges or polylines (1-dimensional) and faces or polygons (2-dimensional). Level 1 adds the possibility to describe real world geographic objects with their characterising properties. The *simple features* in level 1 use the level 0 entities as their geometrical and topological representation and combine them with attributes and relationships. Examples for simple features are signposts, junctions, road elements, and address areas. The features can have attributes such as number of lanes and permissible direction of travel. Relationships between junctions and road elements can be used to model prohibited manoeuvres, e.g. those indicated by *"no left turn"* road signs. In level 2 simple features can be aggregated to describe *complex features*. Examples of complex features are roundabouts and highway junctions.

Depending on the kind of application different levels are used. As guidance through complex junctions requires a high level of detail, level 1 is required for route guidance applications. Level 2 is more appropriate for the calculation of shortest routes, as it is not required to consider the full complexity of how to traverse complex junctions and roundabouts.

2.2.3.4 Applications

Among the fundamental applications in GIS-T are geocoding, route calculation, and map matching.

Geocoding

Geocoding is the process of assigning geographic coordinates, in particular longitude and latitude, to address information. Address information typically includes country, city, street name, and house number. Usually a postal code is added which significantly eases finding the approximate location corresponding to the address. Although address information uniquely defines a certain location, the representation is not very well suited for computerised processing. Geographic coordinates are particular important to determine approximate distances between two points A and B, e.g. between the current position of a vehicle and its destination. For points which are near to each other the Euclidean distance can be calculated by

$$d_{AB} :=\| x - y \|_2= \sqrt{(x_A - x_B)^2 + (y_A - y_B)^2}$$

whith (x_A, y_A) and (x_B, y_B) denoting the coordinates of A and B. The Earth's curvature has to be considered when calculating the distance between points which are far apart, see figure 2.8. Let (λ_A, μ_A) and (λ_B, μ_B) denote longitude and latitude of

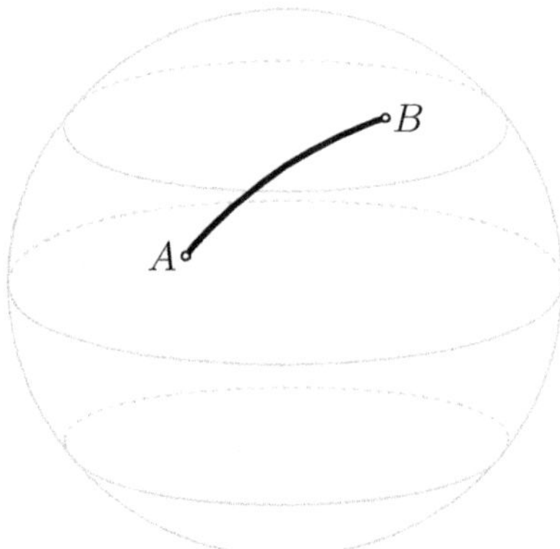

Fig. 2.8: Distance between A and B on the Earth's surface

A and B and let the Earth's diameter of approximately 6370 km be denoted by R. The distance on the Earth's surface can be approximated by

$$d_{AB} := R \cdot \arccos\big(\sin\mu_A \cdot \sin\mu_B + \cos\mu_A \cdot \cos\mu_B \cdot \cos(\lambda_A - \lambda_B)\big).$$

In road transport vehicles cannot travel on the direct line between two points, instead, they travel along the road network. As the distance travelled along the road network is usually longer than the direct line, a fix multiplier[1] is often used to approximate the travel distance.

Route calculations

One of the most important applications in transportation is the calculation of the least cost route from one point A to another point B. Alternatively, the route with the shortest distance, fastest travel time, etc. can be calculated. This problem is known as *shortest path problem* in directed networks, see figure 2.9. The shortest path problem is the problem of finding a path from one point to another minimising the sum of all costs c_{nm} associated to the arcs in the path. The shortest path problem can be solved using the well-known *Dijkstra* algorithm or the A^* algorithm, see Ahuja et al. (1993).

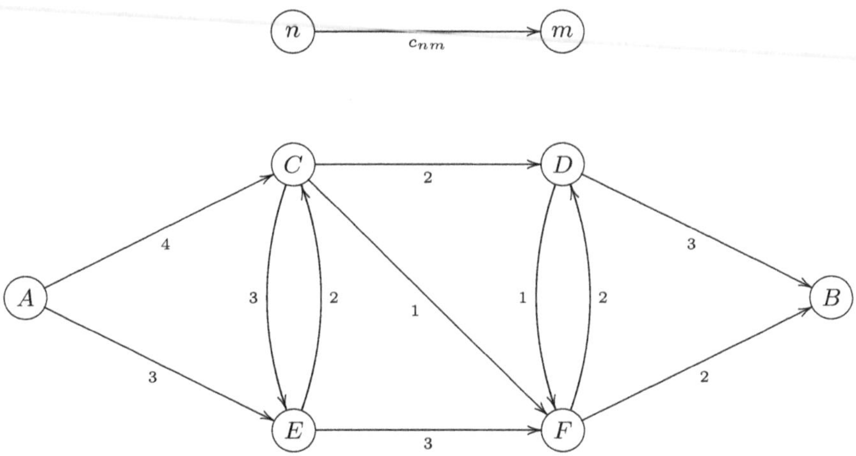

Fig. 2.9: Shortest path problem in a directed network

Map matching

Map matching is the problem of matching a given (set of) estimated location(s) with the corresponding position(s) in the digital representation of the real world, i.e. the digital map. The goal is to match the estimated location of a vehicle with an arc in the network, and then determine the street and the position on the street that corresponds to the vehicle's actual location. Map-matching algorithms are used to reconcile inaccurate locational data with an inaccurate digital map. They can be classified into *point-to-point*, *point-to-curve*, and *curve-to-curve* methods:[2]

[1] This multiplier is typically between 1 and $\sqrt{2}$ (the maximum deviation according to the *Manhattan distance*).

[2] See Bernstein and Kornhauser (1996)

- point-to-point matching: a single vehicle position is matched to the closest node in the road network
- point-to-curve matching: a single vehicle position is matched to the closest point on an arc in the road network
- curve-to-curve matching: a set of vehicle positions is matched to the best "fitting" path in the road network and the vehicle positions are accordingly matched to points on that path

Of course, it is not necessary to determine the distance to every node, arc or path in the road network. Instead, one can first identify those nodes, arcs or paths that are "reasonable" and then only calculate the distance to those nodes, arcs, or paths.

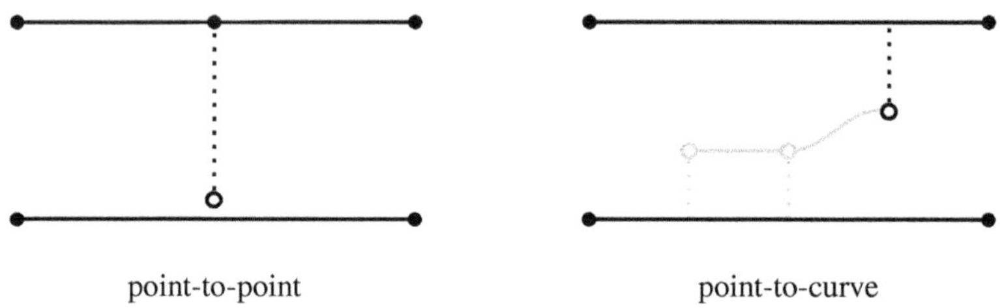

Fig. 2.10: Problems with point-to-point and point-to-curve map matching

As illustrated in figure 2.10, point-to-point map matching encounters problems originating in the way in which the map was digitised, i.e. roads which are digitised in more detail are more likely to be matched. The fact that neither point-to-point nor point-to-curve map matching use any historical information leads to further problems of these methods. Curve-to-curve map matching makes use of historical information and thus, is less sensitive to outliers. Curve-to-curve map matching can be further improved by adding topological information to the algorithm. A thorough discussion of map matching techniques is out of scope of this work and the reader is referred to Czommer (2000), White et al. (2000), and Quddus et al. (2003) for further information.

2.3 Transport telematics

According to Gillette (1988) the *"combinations of computers and telecommunication devices to form new infrastructures are as important to national economies in the twentieth century as the combination of steam engines and carts to form railroads in the nineteenth"*. It is only natural that the combination of transportation and telematics is an important task in the twenty-first century as transportation, computers, and telecommunications are of fundamental importance to every economy. *Transport telematics* concerns the use of telematics with focus on transport organisation, information, and control[1]. The term transport telematics is often used synonymously

[1] See Prognos AG (2001)

Traffic and travel information	• Pre-trip information • On-trip driver information • On-trip public transport information • Personal information services • Route guidance and navigation
Traffic management	• Transportation planning support • Traffic control • Incident management • Demand management • Policing/enforcing traffic regulations • Infrastructure maintenance management
Vehicle-related	• Vision enhancement • Automated vehicle operation • Longitudinal collision avoidance • Lateral collision avoidance • Safety readiness • Pre-crash restraint deployment
Commercial vehicles	• Commercial vehicle pre-clearance • Commercial vehicle administrative processes • Automated roadside safety inspection • Commercial vehicle on-board safety monitoring • Commercial vehicle fleet management
Public transport	• Public transport management • Demand responsive transport management • Shared transport management
Emergency management	• Emergency notification and personal security • Emergency vehicle management • Hazardous materials and incident notification
Electronic payment	• Electronic financial transactions
Safety	• Public travel security • Safety enhancement for vulnerable road users • Intelligent junctions

Fig. 2.11: Fundamental TICS services

to the terms *Intelligent Transportation Systems* (ITS) and *Transport Information and Control Systems* (TICS), however, ITS and TICS are more general and certain applications related to transportation are also termed ITS or TICS applications if they provide transport related use of computer technology - even if no telecommunication is involved. A composite taxonomy of TICS services, as standardised by International Organisation for Standardization (1997), is shown in figure 2.11. A detailed description of all of these applications can be found in PIARC Commitee on Intelligent Transport (1999).

Some transport telematics applications are already widespread and well-known to many transportation professionals and private users. For example, many new (private) cars are equipped with on-board navigation systems considering real-time traffic and travel information. According to a study by Frost & Sullivan[1] the number of commercial vehicles equipped with telematics devices will rise from 75 550 in 2001 to over 5.4 million in 2009. Total market revenues are anticipated to grow from € 169.5 million in 2001 to € 4.7 billion by 2009. As commercial vehicles will be increasingly equipped with telematics devices, the market for telematics services will increase from € 84.3 million in 2001 to just under € 3.2 billion by 2009. It is anticipated that one of the main application of services demanded will focus on logistics and transportation management.

This section gives a short introduction into the major transport telematics applications concerned with commercial vehicle operations.

2.3.1 Traffic and travel information

Traffic and travel information includes information about prevailing and current conditions and regulations concerning the transport infrastructure, points of interest, traffic, and weather. This information is usually categorised into pre-trip and on-trip information. Pre-trip information is used to plan the transport. The transport demand, i.e. the decision whether a transport is done or not, may depend on pre-trip information concerning travel distances and times, tolls, and multi-modal interchange possibilities such as roll-on/roll-off on piggy-back trains or ferries. On-trip information is used to react on the dynamism of transport related issues. Route guidance can be provided dynamically including arrival time estimations considering current traffic and weather conditions. Route guidance instructions are given by acoustic and/or visual turn-by-turn driving instructions. Location based services such as information about nearest truck stops, gas stations, maintenance and repair facilities, etc. can be provided when the vehicle is en-route.

Traffic and travel information are often collected by the transport infrastructure provider and can be disseminated through a variety of media[2]. In North Rhine-Westphalia, for example, traffic forecasts are provided in the Internet, see figure 2.12. Another example is the *Traffic Message Channel* (TMC), a service using the *Radio Data System* (RDS) for sending traffic and travel information using conventional FM

[1] See Frost & Sullivan (2002)

[2] See Kopitz and Marks (1999)

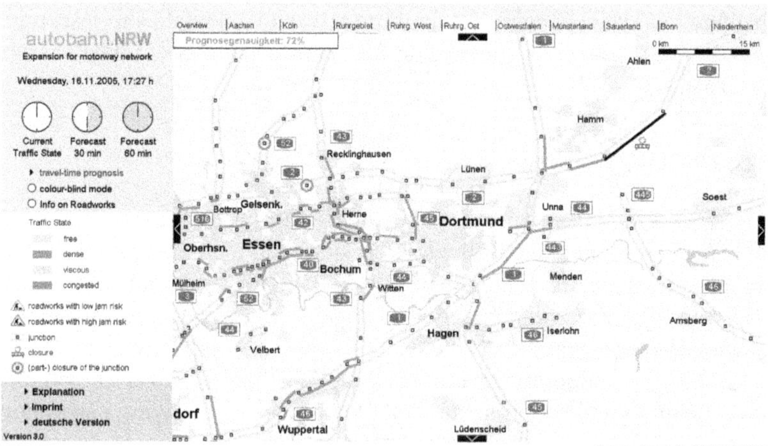

Source: Autobahn NRW (2005)

Fig. 2.12: Traffic forecasts in the Internet

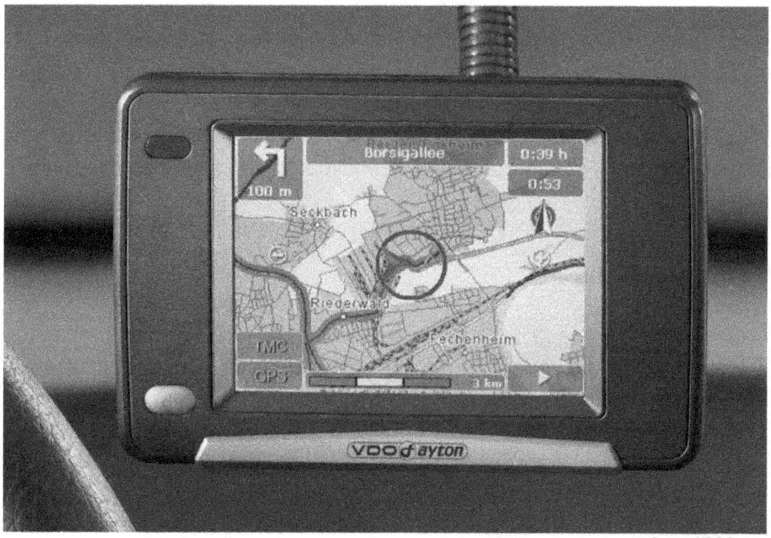

Source: VDO Dayton

Fig. 2.13: On-board navigation system using the TMC

radio. Navigation systems can use this information to calculate shortest routes considering delays due to congestion. Figure 2.13 shows an on-board navigation system using the TMC.

2.3.2 Vehicle-related safety

Vehicle-related telematics applications primarily aim at improving the safety. The European Union initiative to half the number of traffic fatalities until 2010[1] has lead to several projects concerning the use of vehicle-to-vehicle communications based on Dedicated Short Range Communication to improve traffic safety, for example, *Inter-Vehicle Hazard Warning*[2], *FleetNet - Internet on the Road*[3] and *CarTALK 2000*[4]. Vehicle-to-vehicle communications can improve road traffic safety and efficiency by expanding the driver's perception and enabling cooperative driving and platooning[5].

2.3.3 Commercial vehicles

2.3.3.1 Pre-clearance and safety inspections

Control of credentials and other documents, safety status, and weights causes delays for commercial vehicles which increase the cost of transportation. Transport telematics can help in minimising the length and quantity of such stops. Pre-clearance systems enable commercial vehicles to have credentials, other documents, safety status, and weights checked automatically at normal road speeds and without lengthy controls.

2.3.3.2 Fleet telematics

Fleet Telematics Systems (FTS) allow the information exchange between a commercial vehicle fleet and their central authority, i.e. the dispatching office. A FTS typically consists of mobile Vehicle Systems (VS) and a stationary Fleet Communication System (FCS). The FCS may be a stand alone application maintained by the carrier or an internet service running by the supplier of the system. The FCS usually includes a data base in which all vehicle positions and messages are stored. Digital maps are often included which allow to visualise vehicle positions and traces. Figure 2.14 shows an example of such a FCS. Typical components of VS are illustrated in figure 2.15. The communication with the FCS is realised by trunked radio, cellular, or satellite communication. Positioning of vehicles is usually realised by satellite positioning systems and/or dead reckoning using gyroscope and odometer. Usually, the VS is equipped with a simple input device allowing drivers to send predefined

[1] See European Commission (2001)
[2] See DEUFRAKO (2002)
[3] See Hartenstein et al. (2003)
[4] See Reichardt et al. (2002)
[5] See Tsugawa (2005)

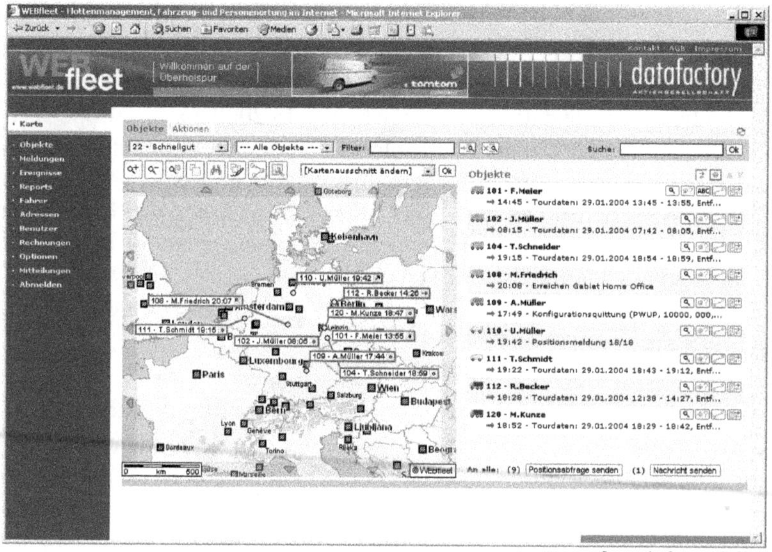

Source: datafactory AG (2005)

Fig. 2.14: Fleet Communication System

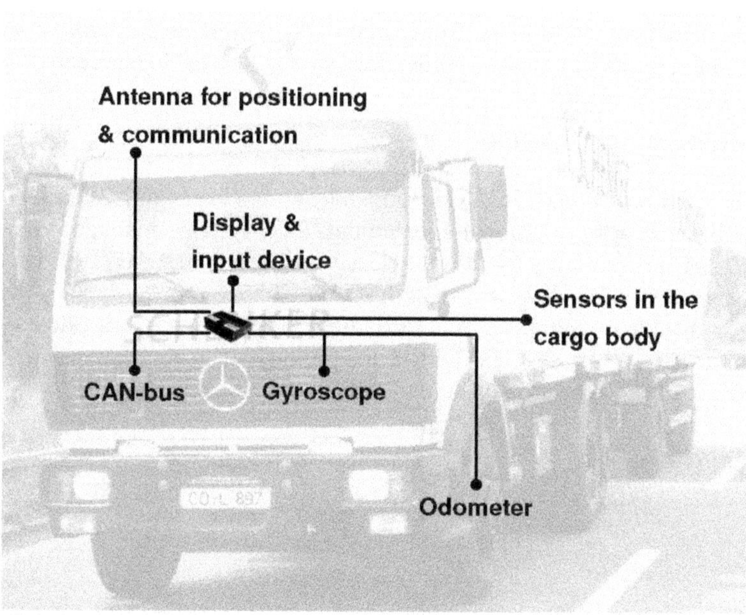

Source (Background image): Schenker AG

Fig. 2.15: Vehicle System

status messages. Drivers may add simple content, e.g. numeric values, but usually cannot enter arbitrary text. Besides of the messages sent by drivers, some VS can also automatically submit messages, e.g. the vehicle's position, data from sensors in the cargo body, or vehicle data from the CAN-bus. In 2002 major European commercial vehicle manufacturers, namely Daimler Chrysler, MAN, Scania, DAF, IVECO, Volvo, and Renault, have agreed to give third parties access to vehicle data using the CAN-bus as a connection. The *Fleet Management Standard* (FMS) is an open standard allowing, dependent on the vehicle equipment, access to vehicle data such as fuel consumption, engine data, or vehicle weight[1].

2.3.4 Emergency management

Telematics systems can provide notifications in case of emergencies, also known as Mayday services. Emergency notifications can be initiated manually, e.g. by the driver pushing a panic button, or automatically, e.g. triggered by airbag, front impact, side impact, and rollover sensors. Vehicle position and the type of damage are transmitted to the service centre. After receiving and verifying the emergency message, the operator at the service centre initiates appropriate measures in co-operation with relevant external organisations, such as police, fire brigades, or medical services. Access to additional information can be provided to the emergency service, e.g. driver-specific information concerning medical data and name of doctor, or data on the nature and condition of hazardous goods. According to Xu (2000) emergency management systems can reduce rescue times by as much as 30 per cent.

2.3.5 Electronic Toll Collection

Electronic Toll Collection (ETC) systems enable drivers to pay tolls automatically on a no-cash basis without stopping at toll stations. ETC systems enable transactions to be undertaken at expressway traffic speed. They are usually based on Dedicated Short Range Communication. However, some ETC systems are also based on satellite positioning and cellular communication techniques. The German TollCollect system[2], for example, uses GPS/GSM and DSRC as a supplement for streets where satellite positioning is likely to be too inaccurate or impossible, e.g. in tunnels.

[1] See FMS-Standard Working Group (2002)
[2] See TollCollect (2005)

3

Commercial vehicle operations

3.1 Introduction

This chapter presents an overview of the development of road freight transport and its impact on commercial vehicle operations. The dramatic changes in the transport industry during the last decades are described and a brief look at the future development of road freight transport is given. The fundamentals of road freight transport are examined focusing on its main characteristics: transportation request, transportation resources, and the transportation services provided. In this chapter strategic, tactical, operational, and real-time management levels are discussed. Operational and real-time tasks, in particular, order and fleet management are discussed in more detail. A case study is presented which will also be referred to in the following chapters.

3.2 Development of road freight transport

During the last decades the road freight industry had to face various new challenges. The most important challenges are caused by globalisation, liberalisation of markets, deregulation in the transport sector, and the increasing commitment to the just-in-time philosophy. The resulting effects on the freight transport industry have been significant in the past and will be considerable in the future.

3.2.1 Globalisation and liberalisation

Today, more goods are transported world wide than ever before. The value of total merchandise exports from all countries of the world increased from € 104 billion in 1960 to € 1625 billion in 1980 to € 5141 billion in the year 2000[1]. As we can see in figure 3.1, the world wide growth of the export volume usually has been magnitudes higher than the growth in the Gross Domestic Product (GDP).

[1] See United Nations Conference on Trade and Development (2004a) (1€ = 1.25 US$)

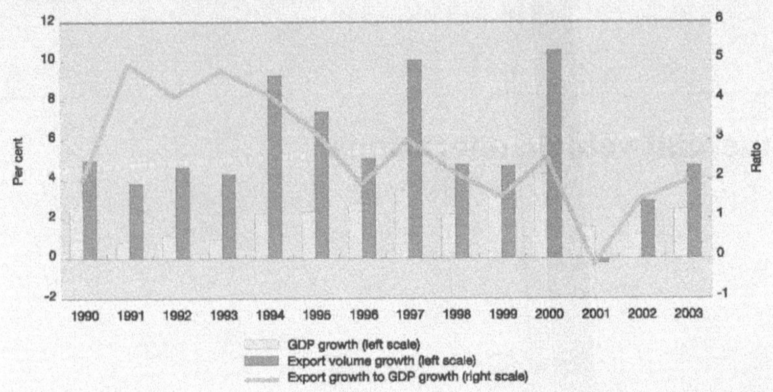

Source: UNCTAD Trade and Development Report 2004

Fig. 3.1: World GDP and export volume growth in the years 1990-2003

Apart from the increase in world wide trade, the establishment and enlargement of the European Single Market has been a cause for increasing intra-trade within the European Union. The value of exports of goods within the EU-25 has grown from € 386 billion in 1980 to € 1294 billion in the year 2000[1].

Due to these effects, the inland freight transportation within the European Union (excluding the 10 member states acceded in 2004) has almost doubled in the years 1970 to 2000[2]. As figure 3.2 indicates, this considerable growth has been almost entirely realised by road transport which has almost tripled in the last 30 years. In other words, the proportion of road transport to total inland transport has grown from about one half in the year 1970 to about three quarters in the year 2000.

According to a study by the European Commission[3], freight transport within the European Union (including the 10 member states acceded in 2004) will increase by about 25% until 2010 and by almost 90% until 2030 compared to the values of 2000. As we can see in figure 3.3, it is assumed that this growth will also be almost entirely realised by road transport.

This development will have considerable effects in future. The transport infrastructure will have problems to cope with such an increase. In some areas the volume of traffic today is already at a critical level and every day 7500 kilometres of European highways are blocked by traffic jams[4].

The freight transport industry accounted for less than one fourth of the total energy demand in liquid fuels in the year 2000 and will account for more than a third of the total energy demand in liquid fuels in the year 2030[5]. This increase in energy demand will result in a higher import dependency of energy sources and oil prices.

[1] See United Nations Conference on Trade and Development (2003) (1€ = 1.25 US$)

[2] See European Commission: Eurostat (2003)

[3] See European Commission: Directorate General for Energy And Transport (2003)

[4] See European Commission (2003a)

[5] See European Commission: Directorate General for Energy And Transport (2003)

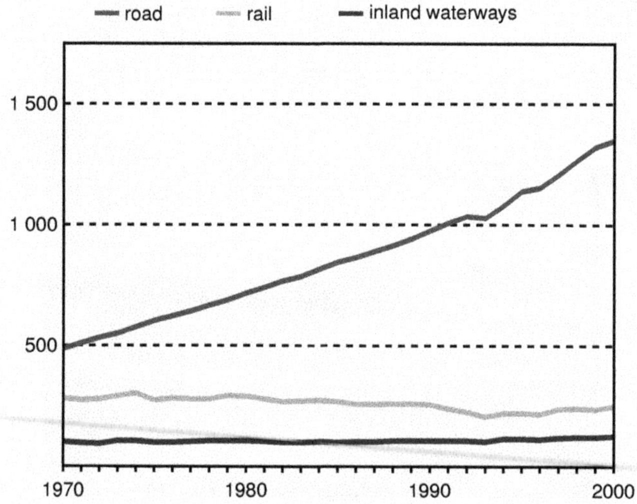

Fig. 3.2: Development of transport by mode (in 1000 tkm) in EU-15 in the years 1970-2000

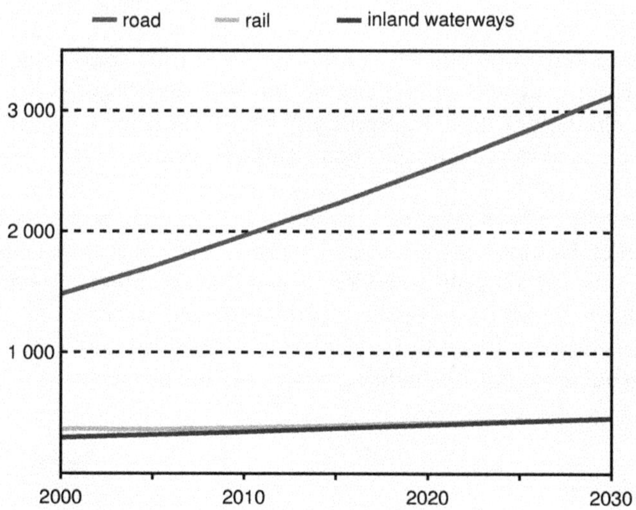

Fig. 3.3: Development of transport by mode (in 1000 tkm) in EU-25 in the years 2000-2030

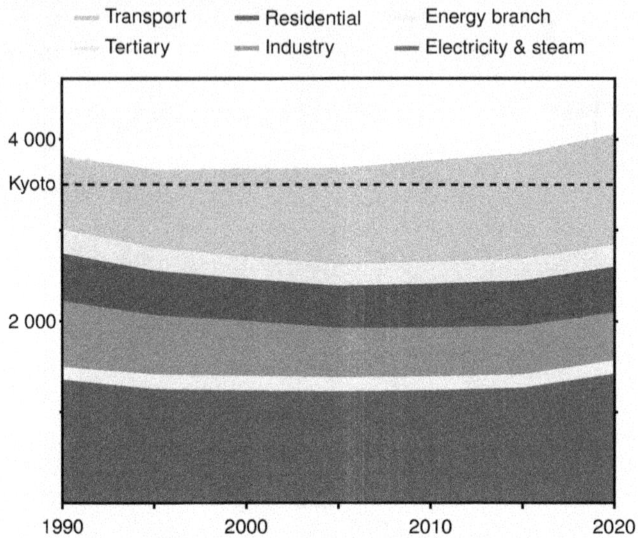

Fig. 3.4: Development of CO_2 emissions by sector (in 1000 t) in EU-25 from 1990-2020. The dashed line indicates the goal of the Kyoto Protocol.

While CO_2 emissions from industry are falling, those from (passenger and freight) transport are rising and will be responsible for 30% of EU emissions of CO_2 by 2010. As a result, the growth of transport will be mainly responsible, if the EU will not achieve their goal of reducing CO_2 emissions according to the Kyoto Protocol, see figure 3.4.

From the motor carriers point of view, the growth in transport activity results in a growing market. However, the developments described above have caused and will cause higher costs and a higher dependency on traffic conditions. Energy prices are likely to rise due to higher prices on the world market and higher taxes. Furthermore, road pricing systems are likely to be increasingly deployed due to economic, ecological, and social limitations in building new roads and the resulting higher level of congestion.

3.2.2 Deregulation

Twenty years ago the European road transport market was heavily regulated[1]. Road haulage between member states of the European Union was only authorised under bilateral agreements, community quota arrangements, European Conference of Ministers of Transport quota arrangements, or for transport exempt from quotas. Carriers involved in international road transport between member states of the European

[1] See Lafontaine and Malaguzzi Valeri (2005)

Union also faced a number of other regulatory constraints that were relaxed between 1985 and 1998. There were lengthy controls at borders up until 1990 when these were eliminated. Furthermore, carriers faced very strict restrictions on international road transport performed by a motor vehicle registered in a third country. So-called *cabotage*, which is defined as transport within a member state performed by a carrier registered in a different country, was completely prohibited. Because these restrictions prevented carriers from picking up as many loads as they might have wanted to along the way, they reduced the utilisation rate of equipment and drivers in international haulage compared to haulage within borders. According to EU regulation 3118/93 and EU regulation 792/94 cabotage operations where finally, after a transitional period, authorised in 1998. Similarly, prohibition against cabotage will be gradually lifted within two to five years for the new member states which accessed the European Union in 2004[1].

Along with the eased access to transportation markets the tariff regulation for border-crossing road-haulage was abandoned and obligatory tariffs for national transport were abolished. This has lead to a dramatic fall in freight transport tariffs which has been estimated between 30 and 40 per cent[2].

The deregulation in the transport sector, in particular, the allowance of cabotage operations, has significantly increased competition between motor carriers. Under the pressure of competition and low tariffs motor carriers have to increase vehicle utilisation and reduce operational costs by adopting better logistics.

3.2.3 Supply chain management and just-in-time practices

Global competition forces manufacturing companies to improve the quality of their products and to reduce their manufacturing costs. Firms are increasingly thinking in terms of competing as part of a supply chain against other supply chains, rather than as a single firm against other individual firms. *Supply chain management* (SCM) refers to the management of materials, information, and funds across the entire supply chain, from suppliers through manufacturing and distribution, to the final consumer[3]. SCM typically involves coordination of information and material flows among multiple firms. Main goals are[4]

- to reduce the time from when a customer orders a product to the final delivery of the product,
- to reduce the inventory of parts, components, and finished goods,
- to increase the accuracy and completeness of filling a customer's order,
- to increase the accuracy and completeness of billing the customer's order, and
- to accelerate the payment for the delivered product.

In order to reduce inventories just-in-time practices are increasingly applied by manufacturers and parts and components are expected to be delivered exactly at the time

[1] See European Commission (2003b)

[2] See Organisation for economic co-operation and development (1997)

[3] See Johnson and Pyke (2001)

[4] See Manheim (1996)

they are needed (just-in-time). One element of just-in-time replenishment is the reduction of order sizes and more frequent requests of deliveries[1]. It is obvious to see that just-in-time practices necessitate close coordination between partners in the supply chain. With reduced inventory buffers and narrow time windows for delivery, any mismatch between supply and demand can result into significant disturbances of manufacturing processes. Thus, manufacturing companies become increasingly dependent on punctual and reliable transportation.

The impacts of just-in-time practices on motor carriers are higher expectations regarding the provided transportation services. Service characteristics like punctuality, reliability and flexibility are becoming more important. Transparency of transportation services is increasingly expected, i.e. shippers and final consumers expect information about the state of order processing and, in particular, estimated arrival times. To face those challenges motor carriers must be linked to the information networks of their partners in the supply chain and must be capable of providing all relevant information.

3.3 Fundamentals of road freight transportation

Commercial vehicle operations are primarily determined by the transportation requests and the transportation resource provided. This section gives an overview of the properties and requirements of shipments to be moved and the properties of vehicles and vehicle equipment. Trucking companies generally specialise in providing different transportation services which will be described in this section: Full-Truckload Trucking, Less-Than-Truckload Trucking, Courier Company Services, and local distribution or collection.

3.3.1 Transportation requests

There is a vast number of different goods which are transported by motor carriers and listing the different types of load that might be carried would be a long task. The various properties and requirements of transportation requests determine whether a load can physically be transported by a vehicle, whether it may be transported according to geographical proximity and logical properties, whether handling operations can be accomplished, and whether the transportation request can be served efficiently.

3.3.1.1 Physical properties

Some cumbersome shipments must be carried by vehicles especially designed to transport those shipments. The majority of shipments, however, can be categorised into certain types which do not require vehicles especially designed for the specific shipment:

- package freight and other non-standardised shipments,

[1] See Srinavasan et al. (1994)

- standardised loading units, i.e. pallets and containers,
- bulk goods,
- and liquids.

Typical examples of package freight are boxes and cartons. They are usually of relatively small size and value, such that they can be consolidated on a vehicle. If freight is not standardised all the measurements have to be known in order to determine whether a set of boxes and cartons fits on a vehicle and whether consolidation with other freight is possible. This may be a quite complex task as some of the shipments may be irregularly shaped and sometimes the exact shape and measurements are not known to the carrier until the pickup location is reached.

The increasing use of standardised loading units has significantly reduced the complexity of transportation. Shippers consolidate freight with the same origin and destination on standardised loading units, i.e. pallets and containers. This eases and speeds up handling, especially if different modes of transport are used.

The most common pallet types are the EUR-pallet and the ISO-pallet. The EUR-pallet is also called a CEN[1] pallet and measures 80 by 120 by 12 cm; the ISO-pallet measures 100 by 120 by 12 cm. Most pallets can easily handle a load of 1000 kilos. The goods are placed on top of the pallet, and can be secured to it by straps or may be wrapped in plastic film. They are often used for deliveries from warehouses to retail establishments.

The most popular container unit is the ISO-container. ISO-containers are 2.44 m (8 ft) wide and either 6.1 m (20 ft) or 12.2 m (40 ft) long. They can be loaded on container ships, railroad cars, and trucks. Terminals and handling equipment are commonly adapted to this standard.

In order to ease handling along the transport chain freight may also be packed in several layers, e.g. a non-standardised shipment may first be palletised and then consolidated in a container, see figure 3.5.

The main physical properties concerning the transport of bulk goods and liquids are volume and weight. Bulk goods are very frequent in agriculture and construction. Examples for the transport of liquids are chemicals, fresh milk, and gasoline.

3.3.1.2 Geographical properties

Pickup, delivery, and service locations

The geographical location of pickup, delivery, and service locations is one of the main characteristics of transportation requests. The most basic transportation requests are requests for picking up some shipment at or delivering some shipment to some location. In online shopping, for example, customers order a certain product, but the specific warehouse where the product is picked up is of no interest to the customer. Another common type of transportation requests is a request for delivering a shipment from one place to another. The shipment has to be collected at its pickup location and must be transported to its delivery location.

[1] Comité Européen de Normalisation

Fig. 3.5: Palletised shipments within a standardised container.

More complex orders request the transportation of multiple shipments between their pickup and delivery locations, for example, when some shipments have to be picked up at a factory and are to be distributed to several other places. Transportation requests may also require the visit of service locations, i.e. locations which are neither pickup nor delivery locations. For example, the transport of chemicals may require that the vehicle visits a cleaning facility after delivering the chemicals.

Route restrictions

Route restrictions may apply for certain shipments, e.g. very heavy shipments or hazardous materials. The use of certain roads may be prohibited and interim locations may be specified to give a detailed definition of the route to be used, see figure 3.6.

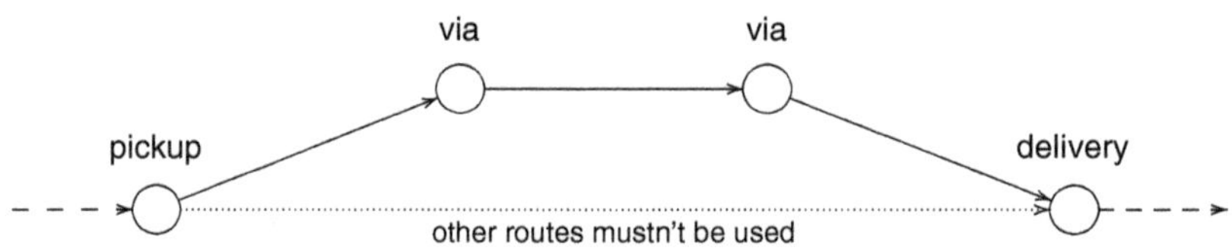

Fig. 3.6: A transportation request with route restrictions

In intermodal transport a part of the trip is done by other modes of transport such as trains or ferry boats. Shippers may request intermodal transport for environmental reasons or to shorten a long trip on land, e.g. to cross the alps. Such requirements result in route restrictions illustrated in figure 3.7.

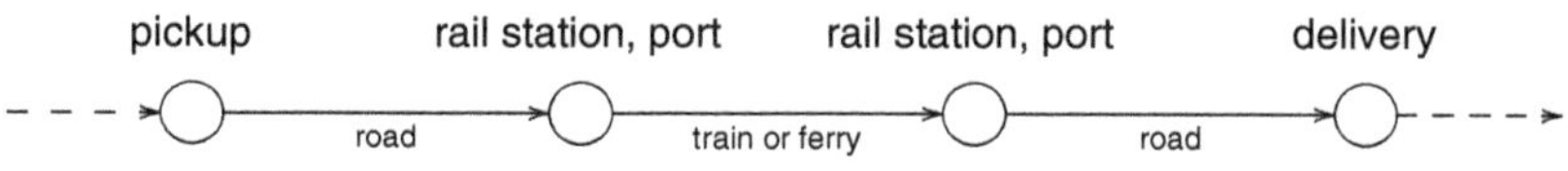

Fig. 3.7: A transportation request with intermodal connections

3.3.1.3 Logical properties

Time windows

Due to the increasing commitment to the just-in-time philosophy most transportation requests are time sensitive. Shippers often request the delivery of shipments until a certain deadline which imposes an upper bound on the delivery time. A lower bound on the delivery time may also be imposed due to business hours or limited storage capability at the destination. At pickup locations a lower bound on the pickup time may be imposed as the shipments which are to be transported may not be ready for transport until a certain time. An upper bound on the pickup time may also be imposed. Time window restrictions may apply to locations between pickup and delivery locations. For example, in piggy-back transport the departure time of ferryboats and trains impose an upper bound on the arrival time at the corresponding locations.

Compatibility constraints

Some loads must not be transported together with certain other loads. For example, different pharmaceutical products may have incompatible admissible temperature ranges. While some chemicals are not hazardous on their own, they might be hazardous if they get in contact with other chemicals. Therefore, load consolidation of such chemicals is prohibited in order to minimise the potential damage in case of accidents. Load consolidation may also be prohibited to prevent damages resulting from additional handling operations.

Some shipments may have specific requirements concerning the vehicles used and their drivers. For example, some chemicals and other hazardous materials, may only be transported if drivers are specially licenced. Some shipments are temperature sensitive and require appropriate equipment in the cargo body. Examples are pharmaceutical products and electronic equipments which must be kept within certain temperature ranges.

3.3.1.4 Handling requirements

Requirements regarding handling operations apply for many transportation requests. While small parcels can easily be loaded and unloaded by the driver, this is not the case for heavy and cumbersome shipments. Pallets often have forklift holes and can be lifted by forklift devices available at pickup and delivery locations. Heavy containers have to be transshipped at specially designed container terminals or with appropriate cranes. Pickup and delivery locations may have ramps such that shipments can be loaded without the need of lifting them. If appropriate facilities at pickup and delivery locations are not provided, vehicles with appropriate equipment are required.

3.3.1.5 Revenue

Each transportation request served contributes a certain revenue to the total yield of the motor carrier. There are different ways of how transportation requests are priced[1]:

- Static pricing: Static prices are standard prices a carrier demands for moving freight between locations. These standard prices are not specific to a contract and are set by the carrier in advance. They are generally the highest prices a carrier will quote.
- Contract pricing: Prices can be fixed on a contract basis. These prices have to reflect the cost of pickup and delivery and moving the freight to its destination. Transportation costs have to reflect the possibility and probability of combining the load with other loads.
- Spot pricing: Spot prices are negotiated according to the current state of the transportation system. They are usually demanded shortly before transportation has to begin and should achieve the yield required to compensate the cost of the decision.

Once it is agreed that a transportation request is served, the price fixed for serving the transportation request is of only minor importance. Usually, some penalty costs apply if the transportation request is not served. Furthermore, external carriers may be subcontracted at a certain price which may be lower than the costs for providing the service by self-operated vehicles.

3.3.2 Transportation resources

Vehicles and equipment operated by a motor carrier have a big impact on the type of shipments which can be transported. Carriers usually concentrate on a certain segment of the transport market and the vehicle fleet is composed accordingly. Physical, geographical, and logical properties as well as the handling equipment provided determine whether a transportation request may be served. The costs for vehicle utilisation determine whether the vehicles can be operated efficiently.

[1] See Powell (2003)

3.3.2.1 Physical properties

Special purpose vehicles may be operated to allow the transportation of certain shipments. Road tankers are used for the carriage of liquids. Tipper trucks and trailers are used in industries that rely on the transport of bulk goods, e.g. agricultural products and waste. Flat-bed trailers are used to carry intermodal containers. Most fleets, however, are operated to transport package freight, small containers, pallets, and other dry freight. Articulated trucks can be equipped with different semi-trailers allowing the transport of various types of shipments. In contrast to articulated trucks, the cargo bodies of vans and rigid trucks are fixed to the vehicle. Articulated trucks offer the greatest payload capacity in terms of gross vehicle weight, but, conversely, may not prove as effective for frequent pickups and deliveries of small shipments.

Source: Europe Container Terminals

Fig. 3.8: An articulated truck with flat bed trailer and containers

Gross vehicle weight, type, and measurements of the cargo body are the main physical properties concerning the transportation of shipments. Serving a transportation request may require several resource classes to be combined in order to get the job done. For example, moving palletised freight may require an articulated truck, a flat-bed trailer, and empty containers, see figure 3.8.

3.3.2.2 Geographical properties

Estimated vehicle position

Usually, each vehicle is stationed at a certain location, e.g. the depot, where the vehicle starts and ends its tour. However, the start and end location of the tour may also be any arbitrary location. Once a vehicle has started its tour, the original starting point of the tour is of minor interest. Instead, the estimation of the current position has to be considered when a tour is modified, see figure 3.9.

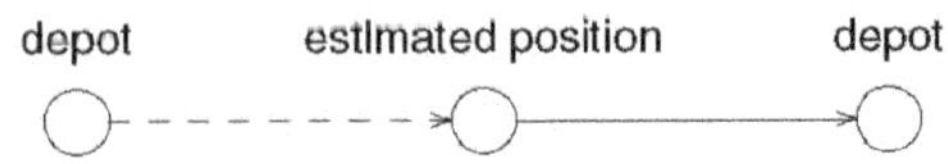

Fig. 3.9: Current position of a vehicle en-route

Vehicles cannot always be diverted from their routes immediately, as delays in communication can occur, or as drivers may have to continue driving until the next exit of a freeway before changing the direction. Thus, dispatchers who plan drivers' tasks have to consider possible vehicle movements until the earliest time the vehicle can be diverted. The position of the vehicle at this time, however, is not known and can only be estimated.

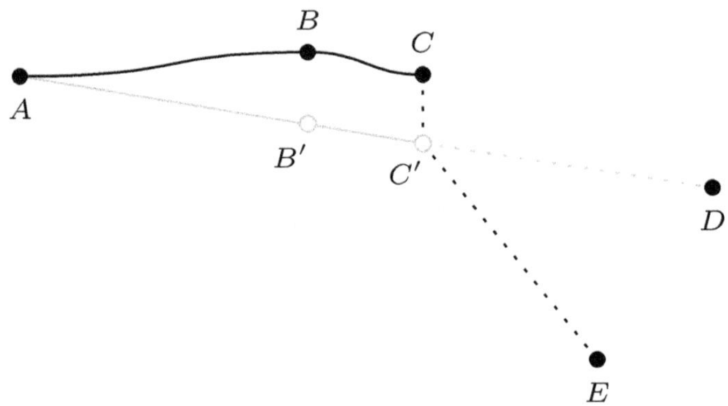

Fig. 3.10: Diversion of a vehicle

Assume that a dispatcher wants to divert a vehicle moving from its last known position A towards its current destination D, see figure 3.10. The current position B (at time t^{now}) of the vehicle is estimated to be at B'. As the modification of tours and the instruction of drivers about their new tasks requires a certain amount of time, planning decisions should not be based on the estimated current position B'. Let t^+ denote the first time the vehicle can be diverted from its current route and let C' denote the estimated position at the time t^+. According to Ichoua et al. (2000), a planning decision made at time t^{now} should be based on (C', t^+). However, the position C' at time t^+ is only an approximation of the actual position C at that time.

Therefore, the inaccuracy of the estimated vehicle position C' at time t^+ must also be considered when diverting a vehicle from its current route towards another point E. Travel times can usually be assumed to obey the *triangle inequality*, i.e. it is never faster to travel from one point to another by visiting an intermediate point as to directly travel from one point to the other. Hence, the additional time for moving from C to E (instead of moving from C' to E) is bounded by the time for moving from C to C'. It is assumed that this time can be bounded by $\tilde{t}$. Planning decisions made at time t^{now}, therefore, should be based on $(C', t^+ + \tilde{t})$.

Route restrictions

Vehicles which are en-route may already have loaded some shipments which still have to be delivered to their destinations. Respective route restrictions must be considered when assigning additional loads to the vehicle as illustrated in figure 3.11.

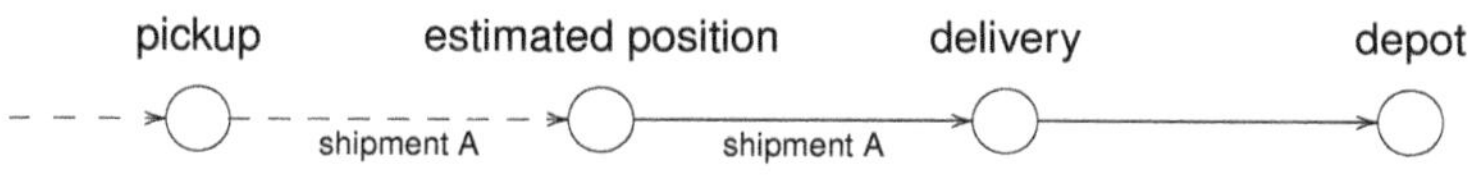

Fig. 3.11: A vehicle en-route with loaded shipments

Other route restrictions occur due to advance booking of train or ferry passages. If a piggy-back transport is booked the resulting route restrictions have to be considered when assigning additional load to the vehicle, see figure 3.12. Similar route restrictions may occur due to maintenance visits at a garage.

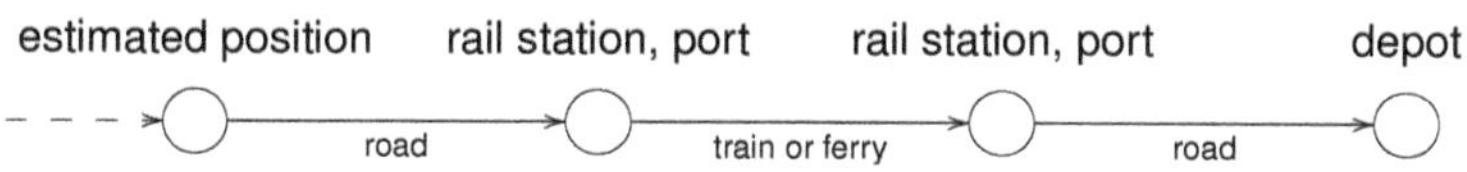

Fig. 3.12: Route restrictions due to a train or ferry passage

3.3.2.3 Logical properties

Drivers' working hours

Since April 2007 drivers' working hours in the European Union are regulated by EC regulation 561/2006[1]. According to the new regulations motor carriers must organise the work of drivers in a way that drivers are able to comply with the regulations, and are made liable for infringements committed by the drivers.

[1] See European Union (2006)

The new regulations demand that a driver shall take an uninterrupted break of not less than 45 minutes after a driving period of four and a half hours, unless he takes a rest period. During a break a driver must not drive or undertake any other work. The accumulated daily driving time between the end of one daily rest period and the beginning of the following daily rest period shall not exceed 9 hours. A daily rest period shall be taken no later than 24 hours after the end of the previous daily rest period. During a daily rest period, which must be at least 11 hours, a driver may freely dispose of his time. Figure 3.13 illustrates an example of driving and rest periods for a vehicle manned by one driver.

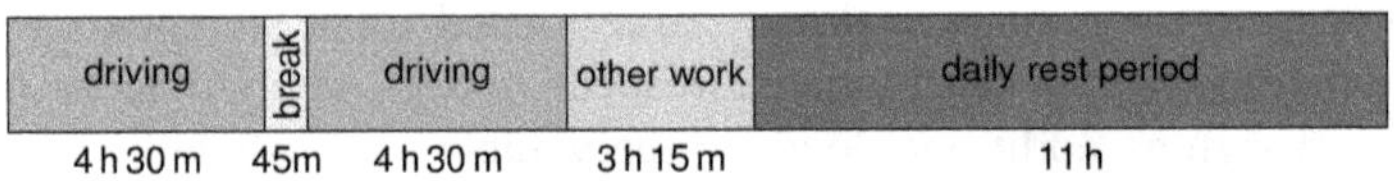

Fig. 3.13: Driving and rest periods for a single manned vehicle

In a multiple manned vehicle, the other driver(s) may take a break on the moving vehicle whilst one driver is driving. Furthermore, the daily rest period in which the vehicle must be stationary may be reduced to 9 hours and shall be taken no later than 30 hours after the end of the previous daily rest period. As can be seen in figure 3.14, travel times for long distance haulage are significantly shorter for vehicles manned by two drivers than for vehicles manned by one driver.

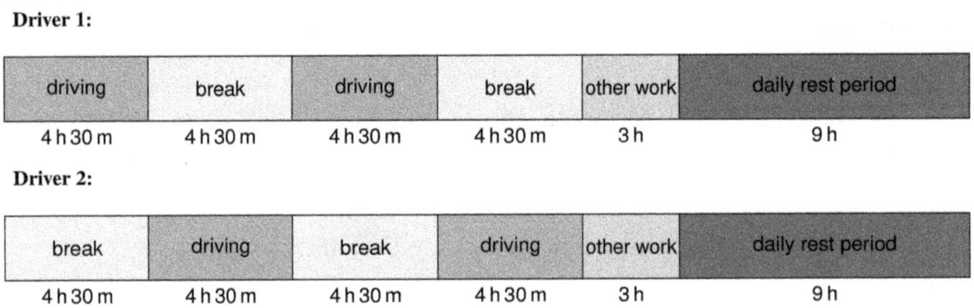

Fig. 3.14: Driving and rest periods for a double manned vehicle

A daily rest period may be taken in two periods, the first of which must be an uninterrupted period of at least 3 hours and the second an uninterrupted period of at least 9 hours, see figure 3.15. Furthermore, a regular break of 45 minutes may be replaced by a break of at least 15 minutes followed by a break of at least 30 minutes.

The daily driving time can be extended to at most 10 hours not more than twice during the week. The daily rest period may be reduced to 9 hours not more than 3

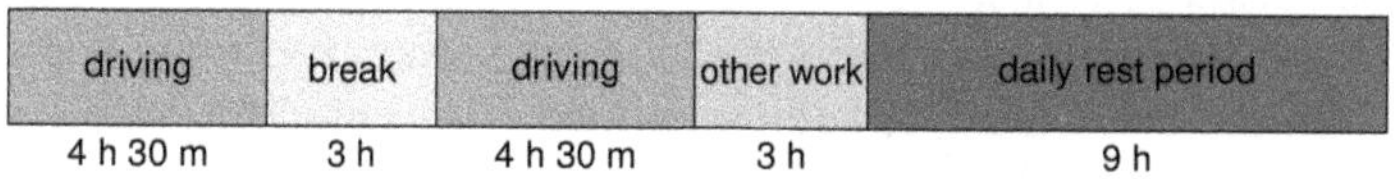

Fig. 3.15: Rest period taken in two parts

times during the week. Figure 3.16 illustrates an example of driving and rest periods with extended daily driving time and a reduced rest period.

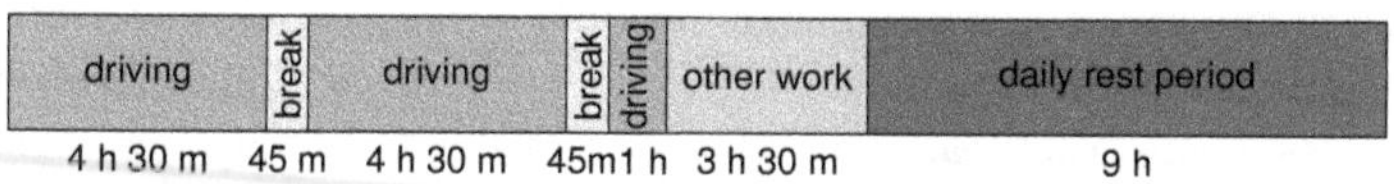

Fig. 3.16: Extended daily driving time and reduced daily rest period

The weekly driving time shall not exceed 56 hours. A weekly rest period shall start no later than 144 hours after the end of the previous weekly rest period. Furthermore, there are regulations regarding weekly rest periods, the maximum weekly working time, and the accumulated driving time during any two consecutive weeks.

Compatibility constraints

Not all transportation requests may be served by all vehicles due to compatibility constraints. Temperature sensitive shipments, for example, may only be assigned to a vehicle if the cargo body is temperature preserving and/or equipped with a refrigerating unit or a heater. Hazardous materials may require that drivers are accordingly licenced. Other licences may allow the transport on certain routes and at certain times, e.g. if driving is restricted on Sundays and at night time.

3.3.2.4 Handling equipment

Some handling activities can be performed by the drivers without any handling equipment. However, it may be required that the vehicle is equipped with hydraulic ramps or cranes, in particular, if no ramps can be used at the pickup and delivery locations. If the shipments must be accessed from the sides, canvas sided cargo bodies or so-called tautliner may be required. Tautliner are special canvas sides which can slide open like curtains and can be opened and closed much faster than normal canvas sided bodies. As rigid boxes can only be opened at the back doors, roller platforms may be required to move the cargo into the interior.

3.3.2.5 Costs

The cost structure of commercial vehicle operations can be fairly elaborate. The degree in which fix costs, like administrative expenses and procurement costs for vehicles, are considered must vary according to the time horizon of decisions. Fixed costs must be considered to ensure long term performance of commercial vehicle operations. On the other hand, if additional load is accepted for a time slot where alternative load is very unlikely to be acquired, fixed costs do not have to be considered. Some costs are directly connected to the transportation processes, for example fuel costs, road tolls, costs for special licences, and costs for handling operations. By EC regulation 3820/85 payments to wage-earning drivers, however, must not be related to distances travelled and/or the amount of goods carried.

3.3.3 Transportation services

Trucking companies usually specialise in the kind of transportation services they provide. In long distance trucking most carriers are specialised either in Full-Truckload Trucking, Less-Than-Truckload Trucking, or Courier Company Services. Another type of service is the local distribution or collection of goods.

3.3.3.1 Full-Truckload Trucking

In Full-Truckload (FTL) Trucking shippers requests an entire truck to move freight from one location to another. After a FTL shipment has been picked up, no other shipments must be loaded until the FTL shipment is delivered to its destination. Motor carriers typically charge an amount based on the distance travelled, rather than on the weight or volume of the contents. When a FTL request is made, the motor carrier has to decide whether to serve the transportation request or not. The decision whether to serve it must consider a possible empty trip to the pickup location and an empty trip from the delivery location to a following pickup location. Furthermore, the probability must be considered that alternative transportation requests with higher profitability will arrive later on.

3.3.3.2 Less-Than-Truckload Trucking

In Less-Than-Truckload (LTL) Trucking shippers requests the transportation of shipments which can vary widely in terms of density and shape and which are too small or otherwise not cost effective to be shipped by an entire vehicle. The carriers seek to consolidate multiple shipments on the vehicle. They have to face the problem of maximising the payload and simultaneously minimising the total mileage. Furthermore, constraints such as accessibility of shipments at their delivery locations and time windows for pickup and delivery must be satisfied. Less-Than-Truckload Trucking usually refers to the transportation of shipments which are larger and heavier than mail-compatible parcels and large volumes and tonnage of freight must be handled.

3.3.3.3 Courier Company Services

Small packages and parcels which are transported by carriers specialised in Courier Company Services must be consolidated on a vehicle in order to be shipped cost effective. In contrast to Less-Than-Truckload Trucking, the delivery of small packages and parcels is characterised by very low resource requirements. Neither capacity nor handling requirements have a big impact on load-to-vehicle assignments.

3.3.3.4 Local distribution or collection

Local distribution usually refers to picking up goods at a central terminal or warehouse, and then delivering the goods to a group of customers. Examples of local collection are the collection of waste from customers in a certain region, or the collection of fresh milk from dairy farms. Local distribution or collection is typically made on a daily cycle tied to business hours. Tours may be fairly regular, e.g. when delivering products to retail outlets, or highly dynamic, e.g. when collecting fresh milk[1].

3.4 Management levels

Management activities of motor carriers can, according to Roy (2001), be classified into strategic, tactical, operational and real-time management levels. Those management levels differ by the short-term and long-term impact they have on future decisions and activities.

3.4.1 Strategic level

Decisions at the strategic management level usually concern a large part of the organisation, have major financial impact, and may have long-term effects. In the motor carrier industry, they typically concern the design of the transportation system:

- the size and mix of vehicle fleet and equipment,
- the type and mix of transportation services offered,
- the territory coverage, including terminal location,
- strategic alliances and cooperations, including the integration of information systems.

Such decisions help to determine the motor carrier's strategic position in the market. They should be revised periodically to respond to changes in the environment. Decisions made at the strategic level constrain the activities and decisions made at the tactical, operational, and real-time levels.

3.4.2 Tactical level

The tactical management level concerns short- or medium-term activities. They typically involve decisions about how to effectively and efficiently use the existing infrastructure and how to organise operations according to the strategic objectives. Activities at the tactical management level include:

[1] See Harrison (2005)

- equipment acquisition and replacement,
- capacity adjustments in response to demand forecasts,
- static pricing and pricing policies for contract and spot pricing,
- acquisition of regularly requested services, including pricing and provisional routing,
- long-term driver to vehicle assignments,
- cost and performance analysis.

In the motor carrier industry, tactical decisions may be performed over a time horizon of a few years when replacing vehicles or a few weeks for the acquisition and provisional routing of regularly requested services. Decisions made at the tactical management level generally influence the activities made at the operational and real-time management level.

3.4.3 Operational level

Operational decisions concern very short-term day-to-day operations. At the operational level, carriers plan current and next day activities in response to daily variations in demand, in equipment, and labour availability as well as short term forecasts. Activities at the operational management level include:

- load acquisition and freight exchange,
- load acceptance,
- short-term driver to vehicle assignments,
- dispatching on a day-to-day basis, including the handout of work plans,
- invoicing.

Decisions made at the operational management level define what the carrier is actually planning to do. They may be influenced by anticipations of future developments, as for example, traffic conditions or the arrival of new transportation requests. Decisions made at the operational level may have to be revised when actual conditions change, i.e. when data becomes known which has not been anticipated.

3.4.4 Real-time level

Commercial vehicle operations underly a variety of external influences which cannot be foreseen. These have to be observed and real-time management decisions have to be made to appropriately react on discrepancies between planned and actual state of the transportation system. Activities at the real-time management level include:

- load acquisition and freight exchange,
- load acceptance,
- real-time dispatching, considering the actual state of the transportation system,
- instructing drivers about their tasks,
- monitoring of transportation processes, including tracking & tracing and arrival time estimations,
- incident management,

- observing the state of order processing.

Activities at the real-time level are dependent of the decisions made at the higher levels and the timely availability of reliable information. They are highly influenced by the inter- and intra-organisational information flow, in particular, the communication between dispatchers and drivers.

3.5 Operational and real-time tasks

Besides of the transportation process itself, operational and real-time tasks of motor carriers include monitoring, control, and planning of transportation processes and order processing, see figure 3.17. The major tasks include load acceptance decisions and dispatching, monitoring of transportation processes, observing the state of order processing, instructing drivers about their tasks, and incident management. These tasks are typically realised by the dispatchers and are the basis for inter-organisational tasks. The main inter-organisational tasks concern load acquisition, freight exchange, and invoicing as well as information exchange with partners in the supply chain e.g. tracking & tracing and proof of delivery.

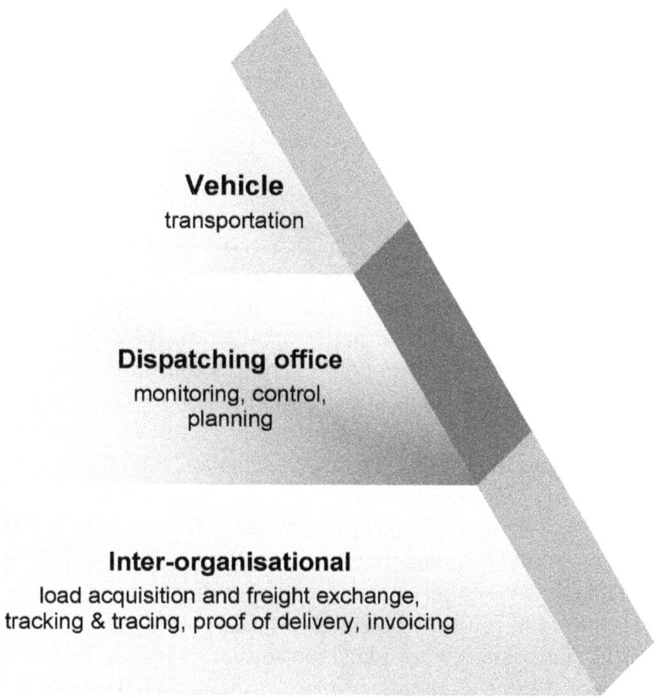

Fig. 3.17: Operational and real-time tasks

Level	Order management	Fleet management
operational	• load acquisition and freight exchange • load acceptance • invoicing	• short-term driver to vehicle assignments • day-to-day dispatching
real-time	• load acquisition and freight exchange • load acceptance • observing the state of order processing	• real-time dispatching • instructing drivers about their tasks • monitoring of transportation processes • incident management

Fig. 3.18: Decisions and activities at the operational and real-time management level

Operational and real-time tasks can substitute or supplement each other and there is considerable overlap between them. Whether a task is realised at the operational or at the real-time level primarily depends on the availability of timely and reliable information and the possibility of communicating with drivers while they are en-route. This section takes a closer look at the tasks performed at the operational and real-time level. As illustrated in figure 3.18, those tasks can be classified into tasks concerning order management and tasks concerning fleet management.

3.5.1 Fleet management

Fleet management concerns all activities needed to monitor, control, and plan transportation processes. Dispatchers have to collect all required information to do so, e.g. by telephoning with drivers. From particular interest are data regarding the current state of the transportation system, i.e. vehicle positions, drivers' working hours, and traffic conditions. This information is required to estimate remaining travel times to the vehicles' next destinations and is a prerequisite for identifying delays. The knowledge of the vehicles' positions is furthermore required to be able to divert vehicles from their current routes towards new destinations. Other information, e.g. odometer information, may be required to plan regular maintenance stops. Traffic and travel information can be used to identify vehicles on congested routes and to divert them to alternative routes without congestion.

Dispatchers have to instruct drivers about their tasks and give them all the information required to fulfil them. This can be done either before drivers start their tour by handing out printed work plans, or while drivers are en-route by using wireless telecommunication techniques. In case of incidents dispatchers have to initiate and coordinate countermeasures. Dispatchers can provide external organisation such as police, fire brigades, emergency services, and breakdown services with essential information e.g. the location of the vehicle. In transportation of hazardous materials,

Transportation request	Vehicle
Physical properties • type of load: container, pallets, bulk goods, liquids, ... • measurements: width, height, length, weight, volume, ...	**Physical properties** • vehicle type: articulated truck, rigid truck with/without trailer, gross vehicle weight • measurements of the cargo body: width, height, length, volume, ...
Geographical properties • pickup, delivery, and service locations • route restrictions	**Geographical properties** • current position of vehicle • route restrictions
Logical properties • time windows • restrictions to load consolidation • temperature sensitivity • hazardous materials • ...	**Logical properties** • travel times • timely availability • driver's working hours • accessibility of consolidated loads • refrigeration unit or heater • licence for transport of hazardous materials • ...
Handling requirements • physical and logical properties of load • equipment at pickup or delivery locations	**Handling equipment** • tautliner sides • hydraulic ramps • cranes • roller surface • ...
Revenue • price the customer is willing to pay • penalty costs if transportation request is not served • costs for subcontracting external carriers	**Costs** • costs for travelled distance • costs for waiting time • handling costs • costs for usage of roads, tunnels, ferries, ...

Fig. 3.19: Characteristics which have to be considered when generating schedules

information concerning the type and state of the loaded shipments can significantly help to take appropriate measures in case of accidents[1].

The most challenging fleet management task is the generation of *schedules*, i.e. the determination of plans indicating which vehicle should visit which pickup, delivery, or service location at what time. The generation of schedules has a considerable impact on the profit a motor carrier can realise and schedules have to be generated

[1] See section 2.3.4

such that the difference between the revenue gained and the cost for vehicle movements is maximised. Figure 3.19 illustrates the main characteristics of transportation requests and vehicles which have to be considered when generating schedules. Depending on the communication possibilities between drivers and dispatchers, dispatching can be made on a day-to-day basis or in real-time. Real-time dispatching can only be done if dispatchers are provided with all actual data concerning the actual state of the transportation system. Furthermore, it must be possible to instruct drivers about new tasks while they are en-route.

3.5.2 Order management

Order management concerns the control and observation of the state of order processing throughout the life cycle of an order, which starts with the acquisition, and ends with the receipt of the payment for the transportation service. Load acquisition may be active or passive. Carriers may simply wait for shippers to call and request the transportation of shipments, or they may identify vehicles with low payload and actively seek loads to be assigned to those vehicles[1].

3.5.2.1 Order management in day-to-day dispatching

Traditionally, there were only few possibilities of communicating with drivers while they were en-route. Instructing drivers was only possible before a driver started his tour or when the driver called the dispatcher, e.g. to report the begin or completion of a transport. Therefore, dispatching was often made on a day-to-day basis and drivers were instructed to serve all orders assigned to their tours by handing out printed work plans before they started their tours. Once the work plans were handed out, no changes to the tour were made.

The state of order processing evolved during the life cycle of an order as illustrated in figure 3.20. After an order is received, the carrier has to decide whether to confirm or reject it. Sometimes this decision has to be made immediately. In other cases, the carrier may wait until a *confirmation deadline*, i.e. the latest time a decision is expected. The load acceptance problem is the problem of effectively choosing the subset of transportation requests to confirm and has significant effect on the profitability and efficiency of the carrier's operations. Load acceptance decisions can be based on an estimate of the cost of providing service and on expectations about future requests[2].

After load acceptance decisions are made, the carrier has to decide whether to serve an order by self-operated vehicles or whether to employ an external carrier to serve the order at a fixed price. The decision whether to serve a transportation requests by self-operated vehicles depends on how the transportation requests can be combined to form tours maximising the profit. In exceptional cases, some of the confirmed orders can neither be assigned to self-operated vehicles nor subcontracted

[1] See Regan et al. (1998), p. 5

[2] See Regan et al. (1998), p. 5

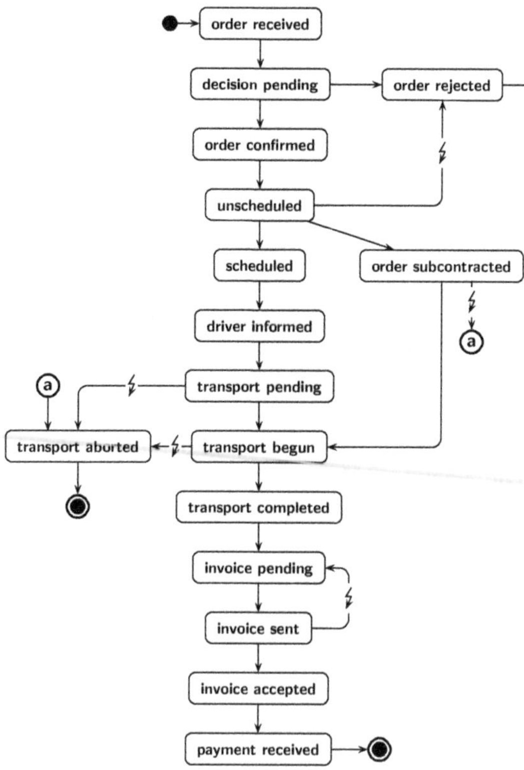

Fig. 3.20: State of order processing in day-to-day dispatching

by external carriers. This may occur due to inadequate policies for load acceptance. In this case the order, although previously confirmed, has to be rejected and usually a penalty fee has to be paid to the shipper. Once an external carrier is subcontracted to serve the order, the actual transportation process is under control of that carrier.

Drivers are instructed to serve all orders assigned to their tours by handing out printed work plans. Once the work plans have been handed out, no changes to the tour are made and the state of order processing is pending for all orders assigned to the tour. When the vehicle picks up the first shipment of the order, the actual transport has begun. The transport is completed when the last shipment has been delivered to its destination. It is assumed that dispatchers are informed when the transport is begun or completed. In exceptional cases, e.g. a vehicle break-down, the transport has to be aborted before the transport has been completed.

After the transport is completed, the carrier has to wait for the way-bill in order to prepare the invoice. The way-bill is usually not sent directly to the accountancy but to the dispatchers, as they are in direct contact with drivers and shippers and much better capable to remove ambiguities in the way-bill[1]. After the correctness of the way-bill has been verified, the way-bill is handed over to the accountancy in order to prepare the invoice. The accountancy sends the invoice to the shipper who verifies and eventually pays the invoice. In the exceptional case that a shipper does not accept the invoice, the dispatcher may again be contacted to explain or correct the way-bill.

3.5.2.2 Order management in real-time dispatching

With the increasing availability of wireless telecommunication techniques the communication possibilities between dispatchers and drivers have improved significantly. Drivers can provide dispatchers with more detailed and frequent information while they are en-route without the need to search for a pay phone. Dispatchers can instruct drivers about new tasks without having to wait until the driver contacts the dispatcher. Real-time dispatching allows the consideration of unexpected incidents and other changes in problem data. New transportation requests for immediate transport can be considered as vehicles can be diverted from their current route in order to serve the new transportation request. In real-time dispatching load acceptance, routing, and transportation processes are performed simultaneously. Schedules can be revised during the transportation processes considering the actual state of the transportation system. The state of order processing evolves during the life cycle of an order as illustrated in figure 3.21.

After a transportation request becomes known, the carrier decides whether to confirm or reject the order and may provisionally schedule it. The scheduling decisions may be revised at any time in order to improve the quality of the schedule or to consider changes in the actual data. Load acceptance decisions, on the other hand, usually cannot be changed. Confirmed orders can either be subcontracted by external carriers or served by self-operated vehicles. In exceptional cases, however, it may neither be possible to schedule a confirmed order nor to subcontract it and the order has to be rejected. In those cases, a penalty fee usually has to be paid to the shipper.

When a confirmed order is scheduled to the tour of a self-operated vehicle, the driver of the vehicle may be instructed to serve the order. To increase flexibility and the freedom of modifying the schedule, this should only be done shortly before the *information deadline*, i.e. the latest possible time the driver has to be informed in order to be able to serve the transportation request. In exceptional cases, a driver may not be able to perform a new task as the information upon which the dispatching decision has been based may have been wrong or incomplete. In those cases, the order is unscheduled from the tour of the vehicle. The order may either be re-assigned to the tour of another vehicle, subcontracted by an external carrier, or may remain unscheduled.

If the driver is instructed before the information deadline, the transport may be suspended. In this case, the order remains scheduled, but the driver will not start the transport unless he is again instructed to do so. Meanwhile the driver may be

[1] See Herrmann et al. (2005)

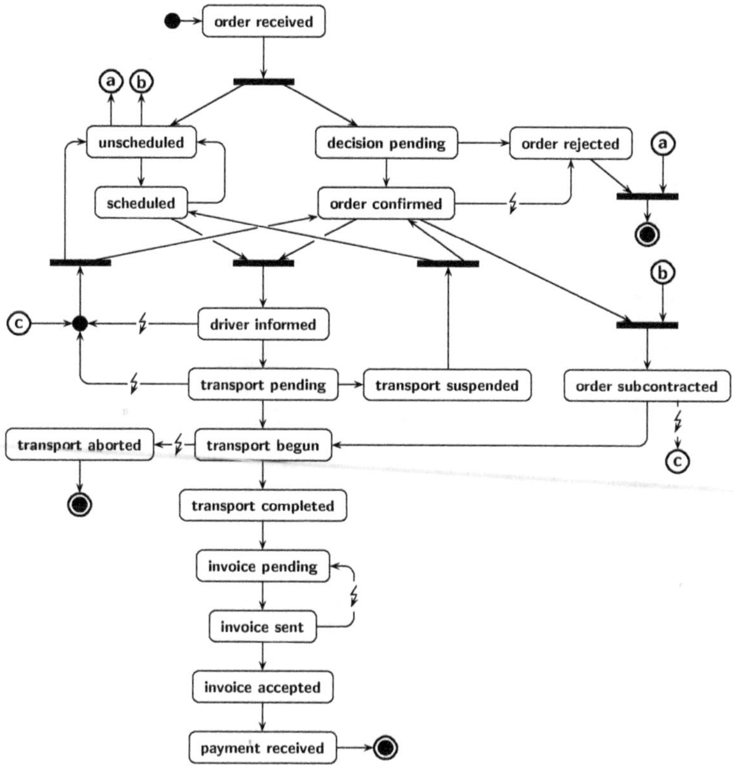

Fig. 3.21: State of order processing in real-time dispatching

instructed to perform other tasks and dispatchers may decide to assign the task to another driver.

In exceptional cases, a pending transport or a transport subcontracted by an external carrier cannot be performed. In those cases, the order is unscheduled and the dispatchers must try to schedule it to another vehicle or to employ another external carrier. Once the transport has begun, the state of order processing evolves analogously as in day-to-day dispatching.

3.6 Case study

This case study[1] examines a motor carrier specialised in so-called *Road Feeder Services* (RFS). RFS are generally used to transport *air freight* to or from a major international airport in order to connect to long haul flights. Apart from some perishable

[1] The case study was motivated by the real case at Georgi Transporte, a German motor carrier specialised in RFS.

and urgent freight, most air freight does not require absolute speed. The reason air transport is chosen for long distance haulage is that customers cannot wait until the next movement by ocean transport to arrive[1]. Furthermore, in short haul transport the relatively slow speed of trucks is often compensated by smaller handling and dwell times[2]. This allows airlines to move air freight overland for considerable distances. Most intra-European air freight is carried by RFS in order to minimise costs[3]. At the airport of Bremen, for example, RFS carry more than 96% of total air freight[4].

From To	Freight Accepted	Days 1234567	Depart	Arrive		Flight Number	Aircraft
From Amsterdam, Netherlands continued							
▸ **London, United Kingdom (Metro) continued**							
	None	12345 7	1200	1205	G	BA8114	EQV
	GE	6	1200	1205	G	BA8114	EQV
27Dec - 03Jan	G	1	1200	1205	G	BA8114	EQV
	None	12345 7	1555	1600	G	BA8116	EQV
	None	12345	1825	1830	G	BA8118	EQV
	G	67	1825	1830	G	BA8118	EQV
18Nov - 30Dec	G	1234	1955	2000	G	BA8120	EQV
	G	5 7	1955	2000	G	BA8120	EQV
04Jan - 24Mar	None	1234	1955	2000	G	BA8120	EQV
	G	123456	1400	0300+1	H	BA9600	RFS
	G	123456	1600	0300+1	G	BA9601	RFS
	G	6	1500	0300+1	H	BA9602	RFS
	G	6	0400	1400	G	BA9603	RFS
	G	6	1000	2200	G	BA9604	RFS
▸ **Manchester, United Kingdom (Manchester)**							
	GEC	123456	0950	1010		BA1625	ER4
	GEC	6	1530	1550		BA1627	ER4
	GEC	12345 7	1650	1710		BA1629	ER4
	GEC	12345 7	2040	2100		BA1631	ER4
▸ **Rotterdam, Netherlands (Rotterdam)**							
	G	12345	1800	2000		BA3600	RFS
	None	12345	1715	1915		BA3608	RFS
	None	123456	1400	1600		BA9600	RFS
	G	12345	1100	1400		BA9807	RFS

Source: British Airways World Cargo (2004)

Fig. 3.22: RFS advertised in a timetable

[1] See Wardman et al. (2002), p. 166
[2] See Becker (1999), p. 56
[3] See Heckmann (2002), p. 42
[4] See Verein Bremer Spediteure e.V. (2005)

RFS are often given their own *flight number* and are advertised in airline timetables. Figure 3.22 shows an excerpt from a timetable of British Airway World Cargo. The RFS with flight number BA9600, for example, is scheduled everyday except Sundays, starts in Amsterdam (14.00 h) and continues via Rotterdam (16.00 h) to London (03.00 h at the next day). Generally, airlines make long-term contracts with motor carriers to provide an agreed number of trucks to perform RFS. Due to variations in actual demand, it may, however, be necessary to acquire additional trucks, or to cancel previously booked trucks, at short notice.

Source: British Airways World Cargo

Fig. 3.23: Contoured air-cargo container.

In order to comply with the special requirements in air transport standardised container and pallets, that fit the configuration of the aircrafts, are used. These standardised container and pallets are referred to as *Unit Load Devices* (ULD). ULD ease the rapid loading and unloading of aircrafts and facilitate the transfer of cargo between aircrafts and trucks. Figure 3.23 shows an example of typical ULD for air freight. The most commonly used ULD have the same ground area measurements and can be categorised in three contour types[1]

- the A-contour with a maximal height of 1.64 m,
- the C-contour with a maximal height of 2.54 m,
- and the D-contour with a maximal height of 3 m.

Air freight generally is time sensitive and has a short life-cycle, is perishable, or needs fast delivery for other reasons. Moreover, connecting flights may impose narrow time windows for RFS and any delay of more than 15 minutes (apart from those

[1] See Heckmann (2002), p. 51

caused by force majeure) results in a contract penalty. Further requirements result from the fact that typical air freight has a high price-to-weight ratio or is perishable, e.g. electronic components or medication. For security reasons double manning may be demanded by the airlines. Additional load may not be permitted in order to prevent damage to the load. Perishable products, such as food or pharmaceutical products, may be temperature sensitive and the freight must be kept within a certain temperature range to maintain an unbroken cold chain.

The motor carrier of this case study operates about 100 trucks and is based near Frankfurt airport, which is the worlds seventh important and Europe's most important cargo hub[1]. The motor carrier provides RFS between Europe's most important airports like Frankfurt, Paris, Amsterdam, and London, as well as less important and remote airports like Helsinki, Lisbon, and Budapest.

Due to the nature of air freight, transportation requests may arrive or may be cancelled at short notice. In many cases new transportation requests must be confirmed while the shipper is on the phone making the request. Schedules must be updated dynamically in order to consider new or cancelled transportation requests. In order to optimise vehicle utilisation dispatchers actively seek for additional load for vehicles with low payload.

The carrier is operating a heterogeneous vehicle fleet composed of articulated trucks and rigid trucks with and without draw bar trailer. The cargo bodies of trucks and trailers are either rigid or canvas sided. Rigid boxes may be required for security reasons when high value shipments are transported. Some of the rigid boxes are equipped with refrigerating units, which may be required for the transport of temperature sensitive goods. As rigid boxes can only be opened at the back doors, roller platforms may be required to move cargo into the interior. Canvas sided bodies can be opened on both sides and the back, enabling loading and unloading from every side. Tautliner are special canvas sides, which can slide open like curtains, and can be opened and closed much faster than normal canvas sided cargo bodies. The capacity of vehicles (trucks including trailer) is expressed in terms of the maximal number and type of ULD that can be loaded. The biggest vehicles have a capacity of 5D, i.e. 5 ULD of contour type D or smaller may be loaded. Trucks are manned by one or two drivers whose driving and working times are regulated by EU social legislation. Travel times and costs depend on the type of vehicle used and the number of drivers.

Several dispatchers collaboratively monitor, control, and plan transportation processes. How this is done will be discussed in the next chapters.

[1] See Airports Council International (2005)

4

Management information systems

4.1 Introduction

Management information systems (MIS) are designed to provide routine information appropriate for planning, organising, and controlling the operations of a functional area in an organisation[1]. This chapter investigates MIS used by motor carriers to perform their tasks at the operational and real-time management level. Many MIS currently used do not have any telematics functionality as, a couple of years ago, only very few commercial vehicles were equipped with telematics devices[2]. The integration of telematics functionality, however, is essential to successfully tackle the challenges motor carriers have to face today[3].

This chapter begins with describing a typical legacy information system without any telematics functionality, focusing on those functions affected by the communication possibilities between drivers and dispatchers. Functionalities provided by fleet telematics systems are described and potentials arising with the use of such systems are identified and classified. It is shown how commercial off-the-shelf fleet telematics systems can be integrated into a typical legacy information system forming a telematics-enabled information system. A Messaging & Fleet Monitoring System is presented which supports the communication between drivers and dispatchers, monitors transportation processes, determines actual data, compares actual data with planned data, and revises planned data in order to consider the actual state of the transportation system. The lack of timely and reliable information used to be a major obstacle for computer-based real-time decision support. This chapter presents a Dynamic Planning System which can be used to provide real-time decision support considering the improved knowledge about the actual state of the transportation system. A transaction control scheme is presented allowing dispatchers, Messaging & Fleet Monitoring System, and Dynamic Planning System to collaboratively and concurrently modify problem data and solution, using different problem knowledge

[1] See Turban et al. (2005)

[2] See section 2.3

[3] See Roy (2001)

and solution techniques. Directions for extending the telematics-enabled information system by additional functionalities provided by electronic freight markets are given. A presentation of the implementation of the Messaging & Fleet Monitoring System and a prototype of the Dynamic Planning System conclude this chapter.

4.2 A typical legacy information system

This section describes a typical legacy information system (LIS) without any telematics functionality. The LIS is the backbone of the motor carrier's information flow and usually well-tested, reliable, and functional, possibly representing years of customisation to consider the unique business requirements of the motor carrier.

4.2.1 System architecture

A typical LIS operated by a motor carrier is illustrated in figure 4.1. The Order & Fleet Management System (OFMS) has the central role concerning the management of orders and vehicles. In the OFMS all data concerning transportation requests and vehicle fleet are stored. In order to distribute information within the information system, the OFMS has bidirectional interfaces allowing other subsystems to connect.

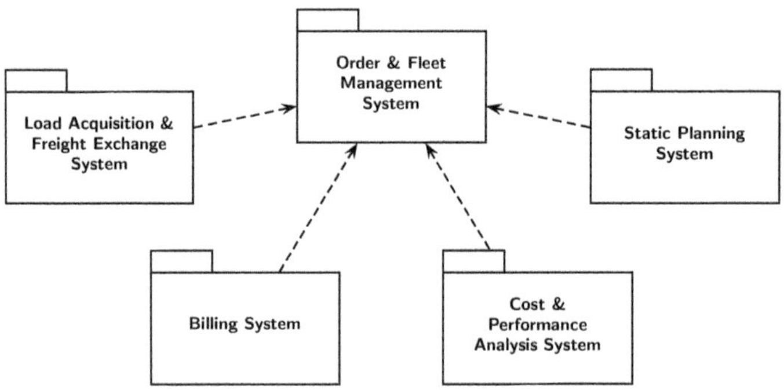

Fig. 4.1: A typical legacy information system

A Load Acquisition & Freight Exchange System (LAFES) can be used to realise the information exchange with shippers who can enter new transportation requests into the system. The LAFES may connect to the information systems of the shippers allowing them to search for appropriate resources and to directly transfer relevant data from their systems into the carrier's information system. After order completion, an invoice has to be generated and sent to the shipper. This tasks is performed with the help of the Billing System (BS) which has access to all relevant data stored in the OFMS, e.g. information about exact arrival and departure times. A Cost &

Performance Analysis System (CPAS) is connected to the OFMS in order to identify unnecessary costs or repeating discrepancies between planned and actual data. Eventually, a Static Planning System (SPS) can be used to generate schedules before transportation begins. Static planning is particularly used when transportation is tied to business hours and planning is done on a day-to-day basis. Transportation requests for the next day are typically collected until a certain deadline and the SPS calculates schedules for the next day assuming that no transportation request arriving after the deadline has to be considered. The optimisation of the static problem is a complex task and may be performed as a batch job during the night.

4.2.2 The Order & Fleet Management System

The Order & Fleet Management System (OFMS) provides graphical user interfaces (GUI) to support dispatchers in monitoring, control, and planning of transportation processes and order processing. Figure 4.2 illustrates the main functions provided to the dispatchers. Dispatchers can observe the state of order processing and have the possibility of updating it e.g. when the transport has begun. They can enter the current vehicle position and its activity into the OFMS. The current activity indicates whether a vehicle is moving or standing due to handling activities, refuelling, rest periods of drivers, maintenance work, or other reasons. To ensure that no information is lost, and to allow the a posteriori identification of responsibilities of incidents, dispatchers have the possibility of entering all information exchanged between drivers and dispatchers into the OFMS. This information can concern the vehicle at a certain time, a trip from one location to another, or certain nodes in the tour of the vehicle. Last but not least, one of the main functions of the OFMS is to support the dispatchers in the generation and modification of schedules.

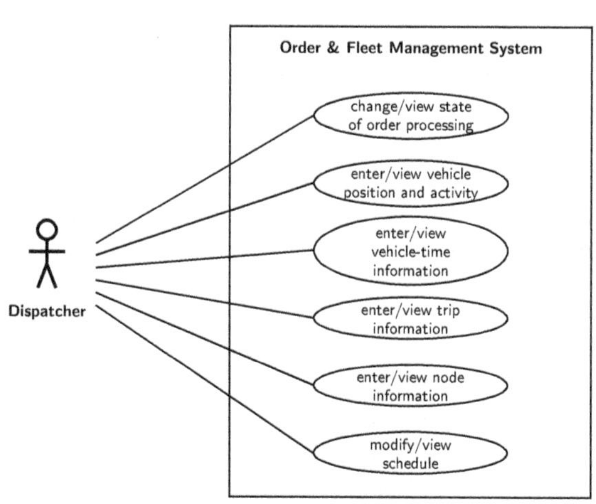

Fig. 4.2: Functions provided by the OFMS

Vehicle fleets are often too big to be effectively managed by a single dispatcher. Motor carriers typically deal with this problem by dividing the covered area into regions of manageable size[1]. Other organisational structures may also apply, but organisational structures are not in the scope of this work. Throughout this work it is assumed that the OFMS is realised according to the client/server paradigm[2] and several dispatchers can collaboratively monitor, control and plan transportation processes.

4.2.3 Communication

The LIS usually does not support the communication between drivers and dispatchers. Instead, mobile telephones or other wireless communication techniques are used to exchange information between drivers and dispatchers as illustrated in figure 4.3.

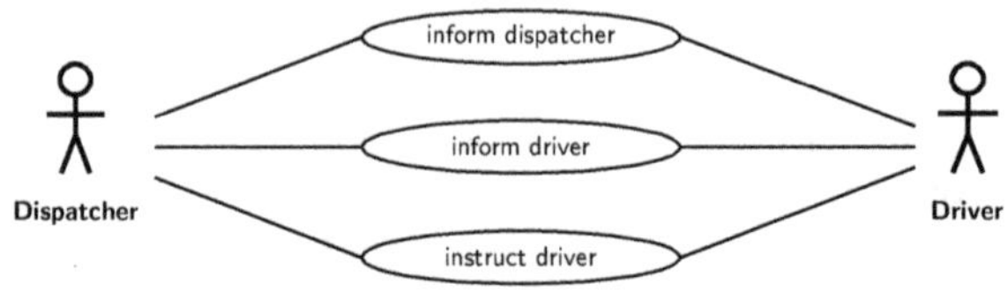

Fig. 4.3: Information exchange between drivers and dispatchers

Drivers are an important source of information for the motor carrier. The information known to the drivers, however, is often not known to the dispatchers. In many cases a driver will call the dispatcher when a task is begun or completed and when major incidents occur. However, drivers cannot report every single detail and cannot report at every single moment. As dispatchers need timely and reliable information, they have to call drivers to actively ask for information whenever the information required is unknown or uncertain. In order to avoid the loss of data and to prevent erroneous planning decisions, dispatchers have to enter all relevant information into the OFMS.

When tasks are assigned to drivers while they are en-route, they must be instructed to perform the tasks. Instructing a driver to serve a transportation request involves supplying the driver with all relevant information, e.g. addresses, dock number, planned arrival times, route restrictions, etc.. As this information may be fairly elaborate, drivers need to write down all the information in a notebook if they are instructed by telephone. Obviously, this bears the risk of misunderstandings and transmission errors. In most cases, a driver will accept a new task, however, in certain cases the driver may refuse to accept it. This may have various reasons. For example, if a dispatcher is not informed about all relevant data, the new task may be infeasible. To indicate whether a driver has successfully been instructed, dispatchers change

[1] See Powell et al. (2002)

[2] See Lewandowski (1998)

the corresponding state of order processing in the OFMS. Dispatchers may also give additional information to drivers, e.g. how to react on unexpected incidents.

4.2.4 Supply chain integration

The foundation of supply chain integration is the inter-organisational integration of information systems which is necessary to coordinate material, informational, and financial flows across a supply chain[1]. Usually, the OFMS is not directly integrated into the information systems of the partners in the supply chain. Instead, supply chain partners can connect to other subsystems, as illustrated in figure 4.4.

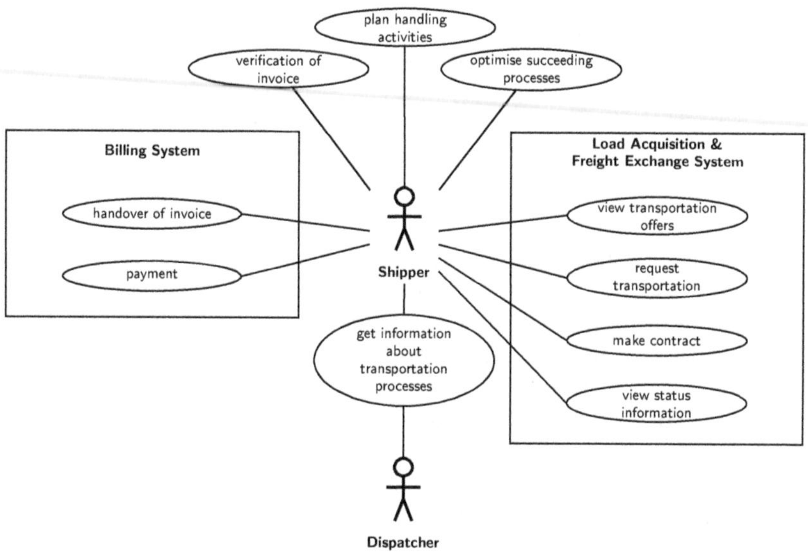

Fig. 4.4: Supply chain integration

The LAFES can be used to view transportation offers and request transportation services. Contracts can be made and the shipper can be supplied with information about the state of order processing. This information, however, is often not sufficient and the shipper may have to call the dispatcher in order to get more sophisticated information, in particular, to be informed about estimated arrival times which are needed to plan handling activities and to optimise succeeding processes. The BS can be used to hand over invoices to the shipper and to facilitate the payment. The shipper has to verify the correctness of the invoice and may need to call the dispatcher in order to get information about transportation processes that is required to clarify obscurities in the invoice.

[1] See Lee and Whang (2001)

4.3 Potentials of telematics

The rapid development of mobile communication and information technology allows the use of telematics to increase the efficiency of commercial vehicle operations. Commercial off-the-shelf Fleet Telematics Systems (FTS) consist of mobile Vehicle Systems (VS) and a stationary Fleet Communication System (FCS) as illustrated in figure 4.5.

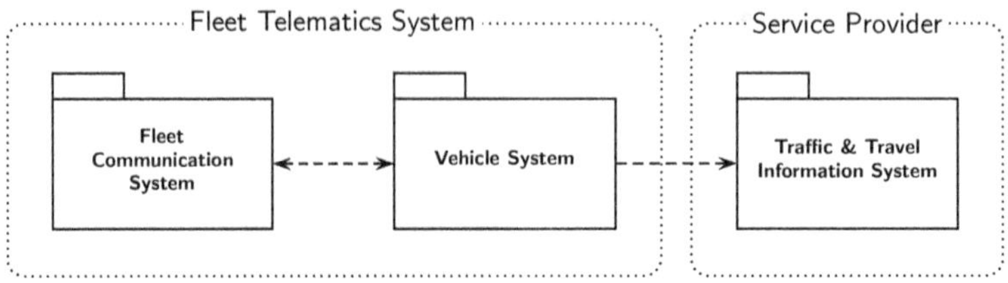

Fig. 4.5: A Fleet Telematics System

The FCS can be a stand alone application maintained by the carrier or an internet service running by the supplier of the system. With the FCS it is possible to exchange data with the VS using wireless communication such as GSM or UMTS. The technical specifications are not essential in this work and it is not differentiated between stand alone applications or internet services or the communication technique used. The VS may be connected to a Traffic & Travel Information System (TTIS) in order to provide route guidance considering actual traffic conditions and estimated travel times.

As illustrated in figure 4.6, dispatchers and drivers can use the FTS to communicate with each other by sending messages. It is assumed that, while dispatchers can send arbitrary messages to the drivers, drivers can only send predefined messages. They may add simple content to the predefined messages, e.g. numeric values, but cannot enter arbitrary text. Although this assumption reduces generality, the increased flexibility accompanied with the possibility of entering arbitrary messages would be an obstacle to automatic analysis of messages sent by drivers. Some VS can automatically submit messages, e.g. the vehicle position or sensor data. Both subsystems of the FTS maintain a history of all transmitted messages allowing dispatchers and drivers to access all transmitted information at any time. The FCS usually includes digital maps and can visualise vehicle positions and traces. The VS may give route guidance in form of acoustic and/or visual turn-by-turn driving instructions.

The potentials arising with the use of a FTS can be classified according to their primary effect:

Availability (A) Information which has not been available before becomes available with the use of telematics.

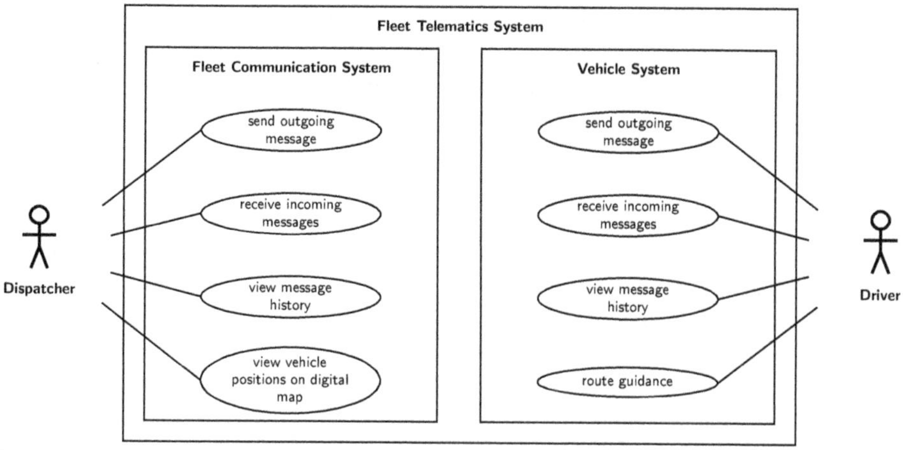

Fig. 4.6: Functions of the FTS

Quality (Q) The quality of information is increased in terms of data integrity and that the information available expresses the actual conditions.

Speed (S) The speed of the intra- and inter-organisational information flow is increased.

The improved information supply in terms of availability, quality, and speed bears further potentials resulting from processing the information:

Usability (U) Information which is available can be readily utilised without the need of manual preprocessing.

Costs (C) Unnecessary costs can be avoided.

Performance (P) The performance of transportation processes can be increased by improved planning possibilities.

Reliability (R) Reliability and punctuality of transportation services can be improved due to better incident detection.

Flexibility (F) The flexibility to quickly react on changed conditions and to modify the schedule can be increased.

Service level (L) The quality of service level can be increased.

Motivation (M) The motivation of dispatchers and drivers can be increased.

The major benefit of FTS is the improved information exchange between dispatchers and drivers. However, the potentials go far beyond the pure improvement in the communication and arise in different fields of application[1]: information exchange, route guidance, tracking & tracing, dispatching, load acquisition and freight exchange, invoicing, and cost and performance analysis.

4.3.1 Information exchange

Drivers are an important source of information for every motor carrier. However, the information known to the drivers is often not known in the dispatching office where it is needed. Furthermore, the information known in the dispatching office may be imprecise or incorrect and often becomes known very late.

Depending on the vehicle system used, messages containing vehicle position and sensor data can automatically be sent to the dispatching office when certain events occur:

- when the dispatching office initiates the transmission of certain information,
- with the receipt of an incoming message (automatic confirmation),
- at fixed times,
- in fixed time intervals,
- after a certain distance travelled,
- when entering or leaving a certain area,
- at the begin or end of vehicle movement,
- with the change of sensor data.

Some vehicle systems allow to configure or reprogramme the logic of those systems while the vehicle is en-route. Those systems allow to control the amount of information transmitted and the cost of communication. In certain situations, some types of information may be very important, and may not be worth the cost of communication in other cases. For example, high value shipments or hazardous materials have special security requirements and intensive tracking is necessary during the transportation process. Empty vehicle movements, on the other hand, do not require intensive tracking and vehicle positions only have to be transmitted once in a while. If the vehicle systems can be configured according to the necessities of the carrier, costs can be controlled and the telematics systems can significantly improve the availability and quality of information. Telematics systems also accelerate the information exchange and thus, can be used to guarantee that all information is available at the time it is needed. Communication costs resulting from transmitting data automatically by a telematics system, are normally lower than those costs resulting from the transmission of the same information via voice communication. However, total communication costs may increase due to the higher amount of information which is transmitted. (A,Q,S,C)

Most vehicle systems are equipped with an input device allowing drivers to send information to the dispatching office. For safety reasons, complex input devices, such as alpha-numeric keyboards, must only be used when the driver is not driving. Therefore, simple input devices are usually preferred as they can be used while driving.

[1] See Lasch (2000)

Many input devices allow drivers to submit predefined messages by pushing certain buttons. Some of them also allow drivers to add numeric values to messages, e.g. a personal identification number (PIN). This PIN can be used for security reasons and to capture working periods of each driver. This allows dispatchers to plan transportation processes considering drivers' working hours. A significant part of the voice communication can be substituted using telematics systems. Drivers can regularly inform the dispatchers about the state and progress of transportation processes at the time the information becomes known. The use of input devices can be much more convenient than voice communication as it is often not practical to call the dispatching office every time some data change. (A,Q,S)

One problem in voice communication is that both sides involved must be capable of communicating at the same time. This, however, is often not the case - neither for drivers nor for dispatchers. Dispatchers cannot always answer telephone calls, e.g. as they may be talking to shippers. Traffic conditions may not allow drivers to use mobile phones, and even if drivers are able to telephone, they may not be able to write down complex information. Thus, drivers may need to interrupt their journey to write down information or to call the dispatcher. This can result in a substantial delay in the information flow. The asynchronous communication provided by telematics systems allows to transmit information at any time without the need of waiting for the other side to be capable of actively accepting incoming information. Furthermore, the communication is not bound to specific persons. As a result, drivers do not have to take care about which dispatcher to call, as any dispatcher can retrieve and process the information once it is stored in the database of the information system. Thus, the processing of information is accelerated. (S,M)

Dispatchers who want to distribute general information to several drivers can be relieved by telematics systems as they usually allow to submit the same message to a group of drivers. Certainly, a dispatcher would be quite busy and unhappy if he would have to make a series of phone calls to inform a large number of drivers. (S,C,M)

Information exchanged by voice communication has to be written down to ensure that no information is lost. Drivers cannot always write down the minutes when they are driving and dispatchers may be busy with more important tasks and cannot immediately key in all relevant data into the information system. Thus, the information is not immediately stored in the OFMS and decisions of other dispatchers may suffer from lack of information. Furthermore, writing down the information requires additional time and effort and is error prone. Telematics systems transmit digital information which can be automatically archived in the vehicle system and the information system in the dispatching office. There is no need for manually entering the information into the systems and all data can be accessed at any time. This relieves dispatchers and drivers.The electronic communication guarantees that all the information received is identical to the information sent. Thus, transmission errors and misunderstandings are avoided. (Q,S,C,M,U)

Generally, most of the data a driver needs for fulfilling his tasks are available in the OFMS. These data can automatically be retrieved from the OFMS and sent to the driver without the need of manually typing in all the data into the telematics system.

The instruction of a driver about his next task can be made completely automatic, however, usually the dispatchers will manually initiate the transmission of messages to the drivers. Message templates can automatically be provided considering all relevant data stored in the OFMS. (Q,S,C,M,U)

With the use of a telematics system telephone calls can be reduced in terms of quantity and duration. However, telephone calls cannot be fully replaced by electronic communication as the flexibility of voice communication is higher. In particular, dialogue-like conversations are impossible using electronic communication. Furthermore, most wireless communication technologies have limited bandwidth and unstructured information can be better exchanged by voice communication. Consequently, telematics systems should not be used to replace voice communication, but to support communication.

4.3.2 Route guidance

Today, many modern cars are equipped with navigation systems. Those systems calculate the optimal route (shortest or fastest) from the current position to a desired destination. The systems give visual and/or acoustic *turn-by-turn* driving instructions which are derived on the basis of the determined optimal route. If the driver does not follow the recommended driving instruction at some location, the system calculates a new optimal route from this location to the desired destination. So-called dual-mode or hybrid route guidance systems integrate real-time information to allow dynamic route guidance taking into account the actual traffic conditions[1]. This information is sent from a traffic service centre through FM broadcasting using RDS-TMC. Navigation systems help reducing travel times and improve the driver's orientation. Furthermore, safety is increased as drivers do not get distracted by finding their way in printed maps. Today, those systems are focused on the needs of private users. Despite of the benefits of (private purpose) navigation systems, those system have the disadvantage that they are not always well suited for commercial vehicle operations. Important information for large commercial vehicles are often not included in the digital maps. Especially for heavy vehicles, optimal routes have to be calculated taking into account various other aspects, e.g. the height of bridges and tunnels, road tolls, no through roads, etc.. In contrast to private usage, commercial vehicle movements are usually planned by a central authority, the dispatchers. Route guidance systems for commercial vehicle operations should therefore, allow to submit the planned route from the dispatching office, where the planning is done, to the vehicle, where the transport is done. If the planned tour is synchronised between dispatching office and the route guidance system deviations can be easily detected within the vehicle. The vehicle system can send a message to the dispatching office to inform the dispatchers about the deviation. The new route which is calculated can then also be synchronised between dispatching office and the route guidance system. Unnecessary mileage can be avoided due to route guidance and early identification of deviation from the planned routes. (A,Q,C,R)

[1] See Xu (2000)

4.3.3 Tracking & tracing

Tracking describes the continuous determination of position and state of shipments and vehicles, whereas *tracing* describes the retrospective identification of the trace of shipments and vehicles[1]. Without the use of a telematics system tracking & tracing functionalities can only be realised at the transshipment points, but not while the vehicle is moving.

Vehicle systems equipped with a positioning system connected to a communication device can regularly send the vehicle's position to the dispatching office. The hereby obtained trace of the vehicle can be be matched with the planned route and it can be identified whether a vehicle significantly deviates from the planned route or whether a trip is significantly delayed. If geographic or timely discrepancies are detected countermeasures can be initiated. (A,S,U,R)

With the help of telematics systems the arrival at and the departure from a handling location can be identified. Exact arrival and departure times can be determined. Further information can be put in context to the handling location e.g. the start and end time of handling activities and the shipments which are loaded or unloaded. The exact knowledge of information related to handling activities can be used to provide a proof of delivery and can be considered when preparing invoices. (A,S,R,U,L)

Throughout the transportation processes, sensors in the cargo body can measure the temperature and messages can automatically be sent if the temperature leaves the required temperature range. This information can be used to initiate countermeasures or to proof that throughout the transportation chain temperature sensitive shipments have been kept at the required temperature. (A,S,R,U,L)

Precise costs of transportation can be proven to the shipper including all unplanned costs, for example, the costs for using a tolled road or the costs for unnecessary waiting times at the shippers location. (A,S,U,L)

Security of high value goods can be increased as vehicles and shipments can be localised at any time. Furthermore, sensors can detect certain incidents, for example, when the cargo body is opened. If this happens at a location where no handling activities are planned, an alarm message can be sent automatically. If an alarm message is received in the dispatching office, an emergency service can be automatically informed. The vehicle positions can be continuously handed over to the emergency service enabling them to quickly find or follow the vehicle. (A,S,R,L)

Drivers' working hours are regulated by EU social legislation. If the dispatchers do not have the full knowledge about driving and rest periods, decisions made by the dispatchers can result into conflicting instructions resulting into delays or violation of drivers' working rules. The knowledge about work and rest periods leads to better schedules, better safety on roads, and better satisfaction of drivers and dispatchers. (A,U,R,M)

Telematics systems can be used to identify whether a vehicle is moving or not. The reasons why a vehicle is not moving can be identified and unnecessary idle times of vehicles can be detected and avoided. (A,U,C)

---

[1] See van Dorp (2002)

The knowledge about the current position and state of the vehicles and shipments can be used to calculate arrival time estimations which can be provided to shippers. Shippers need this information to plan handling activities and optimise succeeding tasks. (U,L)

Eventually, all data identified can be stored in the information system allowing to trace movements of vehicles and shipments. If the information systems of all freight forwarder involved are connected, tracking & tracing functionalities can be provided throughout the whole transportation chain - not necessarily restricted to part performed by the motor carrier. (A,U,L)

4.3.4 Dispatching

With tracking & tracing of vehicles and shipments, the knowledge about the current state of the transportation system can be significantly improved. The continuous supply of timely and reliable information allows the modification of schedules considering the actual state of the transportation system. Hence, the flexibility in schedule generation is improved and carriers no longer have to hand out printed work plans when drivers begin their tour. Drivers only need to be informed until the information deadline of the task[1]. By waiting until this information deadline much more time is available for modifying plans which may become necessary when new transportation requests arrive or other changes become known. The improved knowledge about the actual state of the transportation system, helps in increasing the quality of schedules and the reliability of transportation processes. (P,F,R)

The lack of timely and reliable information is the biggest problem for computer-based real-time decision support[2]. The reason for this is the high frequency with which the real-world situation changes and the tremendous effort which is needed when information has to be manually entered into a software system. As a result, automatically generated schedules often cannot comply with the requirements of real-life problems. The effort needed to manually adjust automatically generated schedules is tremendous and, after manual adjustment, the quality of those schedules is not necessarily much better than a manually generated schedule. Due to tracking & tracing of vehicles and shipments the information gap between drivers and dispatchers and the information system can be significantly reduced. This allows the use of dynamic planning systems to optimise schedules considering the actual state of the transportation system. (P,F)

4.3.5 Load acquisition and freight exchange

Motor carriers seek to minimise empty mileage and to maximise the payload. Tracking & tracing allows the carrier to optimise the schedules considering all known transportation requests. As a result of the improved information supply, vehicles with low capacity utilisation can be identified easier and earlier. Thus, carriers can

[1] See section 3.5.2.2

[2] See Powell et al. (2002)

improve their performance by actively searching for additional load for vehicles with low capacity utilisation. (P,F)

When incidents are detected, the carrier can reconsider whether already confirmed transportation requests should be served by self-operated vehicles or not. The carrier may want to remove some transportation requests from the tours to guarantee punctuality of the remaining orders. If no self-operated vehicle can efficiently serve the unscheduled transportation request, the carrier has to reject the order unless additional transport resources can be acquired. The carrier can search for additional transportation resources in electronic freight markets. If successful, external carriers can be subcontracted in order to perform the transportation. (P,R)

4.3.6 Invoicing

All necessary information for preparing an invoice is written into the way-bill by the drivers who return the way-bill to the dispatching office after order completion. As noted by Herrmann et al. (2005), the time needed from order completion to begin of preparing the invoice can be as high as two weeks. With the use of telematics systems all relevant information can be sent to the dispatching office when the information becomes known and before the way-bill is returned. Possible obscurities in the data can be resolved easier as the corresponding occurrences are more present to the persons involved. A *proof of delivery* can be given directly after order completion and a detailed invoice can be prepared without having to wait until the way-bill is returned. This speeds up the billing process, eases the verification of invoices, and results in an earlier payment of transportation services. (C,U,S,L)

4.3.7 Cost and performance analysis

In order to support the analysis of the carrier's costs and performance, telematics systems can be used to provide comprehensive data. Repeated discrepancies between planned and actual data as well as their causes can be identified. For example, repeatedly long travel times of all vehicles on a specific route can be used to modify travel time estimates. Repeatedly high fuel consumption of a single vehicle may indicate engine problems or uneconomic driving behaviour of the driver. Repeatedly long waiting times at certain handling locations can be identified and agreements with certain shippers can be reconsidered. Expenses caused by technical problems, uneconomic driver behaviour, and bad planning data can be identified and avoided. (A,C,U,P)

4.4 The telematics-enabled information system

The potentials discussed in the previous section can be exploited by using telematics. Commercial off-the-shelf FTS, however, usually cannot provide any functionalities beyond an information exchange, route guidance, and the visualisation of vehicle positions on a digital map. Section 4.2 described a typical legacy information system (LIS) without any telematics functionality. If the commercial off-the-shelf FTS is not integrated into the LIS the information available is distributed throughout the carrier's IT infrastructure, as information sent electronically by the vehicle systems is only available in the FTS, while information obtained by talking to drivers is stored in the LIS. Manually transferring the information from the FTS into the LIS results in additional effort and is error prone. It leads to redundancies and increases the risk of operational errors as data integrity is not guaranteed. Furthermore, the FTS is a rich source of information and it is impossible to manually transfer and analyse all relevant information to the LIS.

As the LIS is the backbone of the motor carrier's information flow, it can be seen as mission critical. It is usually well-tested, reliable, and functional, and users are very familiar with it. A replacement of the LIS is usually not possible as replacing the LIS not only involves the direct financial cost of software development, but also the expense of rediscovering the accumulated knowledge about business rules and processes. The high risk of failure and the effort for training the personnel are further obstacles to a replacement of the LIS.

This section describes how a typical LIS of a motor carrier can be modernised by integrating a commercial off-the-shelf FTS. It is examined how the communication between drivers and dispatchers can be supported by a Messaging & Fleet Monitoring System, and how the Messaging & Fleet Monitoring System can monitor transportation processes by analysing messages sent by the vehicle systems. The telematics-enabled information system improves the knowledge about the actual state of the transportation system, and provides the basis for computer-based real-time decision support. This section presents a Dynamic Planning System (DPS) which can provide real-time decision support and gives directions for extending the telematics-enabled information system by additional functionalities provided by electronic freight markets are given.

4.4.1 System architecture

Usually, neither the commercial off-the-shelf FTS nor the LIS can be modified easily. It is however assumed, that all relevant data is accessible from outside the systems. In other words, it is assumed that interfaces exist that allow a system to carry out the same tasks as the human users can. The FTS can be integrated into the LIS by the deployment of a Messaging & Fleet Monitoring System (MFMS) as illustrated in figure 4.7 and described in Goel and Gruhn (2005). The MFMS connects to the OFMS to access information about all planned tasks. In order to instruct drivers about their tasks, the MFMS sends messages to the vehicles via the FCS. Furthermore, it retrieves actual data provided by the FCS and the TTIS. This actual data is analysed and compared with the planned data. Eventually, the MFMS revises the planned data stored in the OFMS according to the actual state of the transportation system.

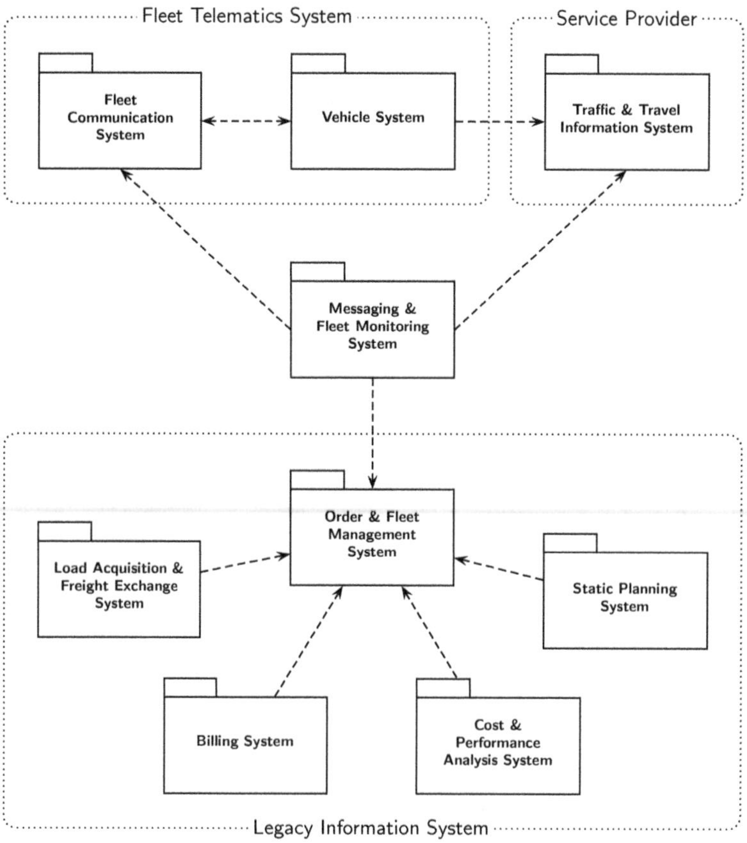

Fig. 4.7: The telematics-enabled information system

4.4.2 The Messaging & Fleet Monitoring System

The main tasks of the Messaging & Fleet Monitoring System are to support the communication between dispatchers and drivers, and to monitor transportation processes. In order to do so the MFMS builds a bridge between the actual data provided by the FCS and the TTIS, and the planned data provided by the OFMS. As neither the LIS nor the FTS know about each other, the MFMS provides the functions illustrated in figure 4.8.

In order to support the communication between dispatchers and drivers, the MFMS provides user interfaces to write messages and to instruct drivers about new tasks. These user interfaces provide message templates incorporating data stored in the OFMS, e.g. the address of handling locations and the planned arrival times. Furthermore, the MFMS can display notifications to the dispatchers, e.g. if disturbances in transportation processes are identified.

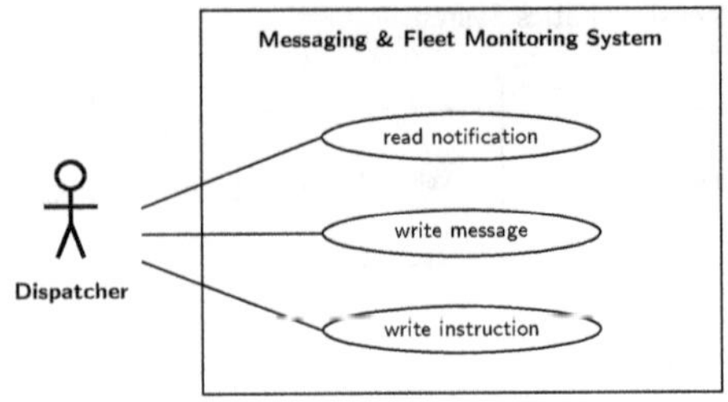

Fig. 4.8: Functions provided by the MFMS

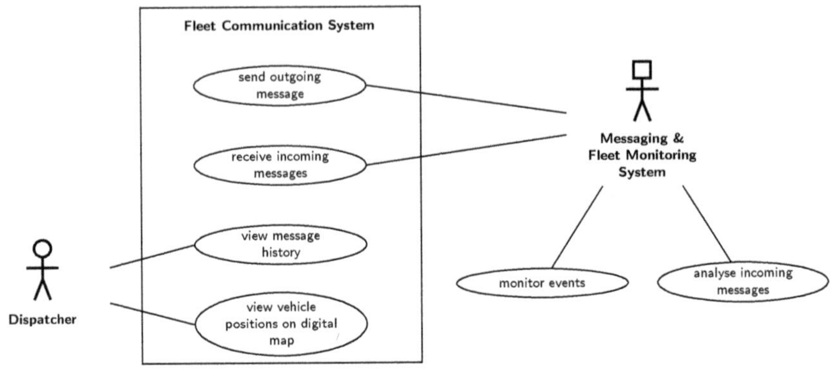

Fig. 4.9: The entire communication is realised using the MFMS

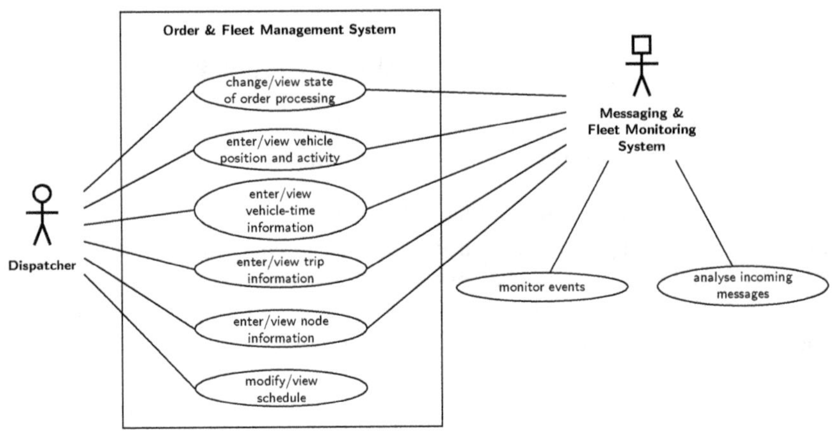

Fig. 4.10: The MFMS takes over the role of the dispatchers

As the MFMS provides user interfaces to allow dispatchers to write messages and instruct drivers, the FCS is no longer used by the dispatchers. Instead, the MFMS hands over all messages and instructions to the FCS, as illustrated in figure 4.9. In order to determine actual data, the MFMS analyses all incoming messages and monitors all planned events. The actual data is compared with the planned data and revised planned data is stored in the OFMS to consider the actual state of the transportation system. If ambiguities or problems are identified, the MFMS notifies dispatchers. Only messages which cannot be unambiguously interpreted by the MFMS have to be analysed by the dispatchers. In order to access and modify data stored in the OFMS, the MFMS takes over the role of the dispatchers as illustrated in figure 4.10.

4.4.2.1 Communication

The communication technique provided by a FTS is asynchronous and has the advantage that dispatchers can transmit messages and instructions to drivers at any time, regardless of whether the drivers are in the vehicle or not, or are otherwise prevented from answering a phone call and writing down the information. However, as the communication is asynchronous, dispatchers do not immediately know whether a driver has read and understood the messages sent to him. Therefore, drivers are usually instructed to send a confirmation to each message sent to them.

The MFMS supports the information flow as illustrated in figure 4.11. A message, which is composed by the dispatcher, is sent to the driver by the MFMS and stored in the OFMS. The MFMS monitors all incoming messages, checks whether a confirmation was sent and stores it in the OFMS. If the confirmation is not sent until a certain time limit after the message was sent to the driver, the MFMS detects that a planned event has not occurred and notifies the dispatchers. If the confirmation was sent too late, the dispatchers are also warned as they may already have assumed that the message was not read by the driver.

Most of the data a driver needs for fulfilling his tasks are available in the OFMS. The MFMS can use these data to support dispatchers in instructing drivers to perform a task as illustrated in figure 4.12. First, the dispatcher changes the state order processing to indicate that the driver is instructed. This is observed by the MFMS which generates a message template with all the relevant data. The dispatcher can modify this message before it is sent. The message is then sent to the vehicle by the MFMS and stored in the OFMS. When a reply is received, it is stored in the OFMS and analysed to determine whether the driver accepts the task. If the task is rejected, the MFMS changes the state of order processing to indicate that no driver has yet been successfully instructed to serve the transportation request. A warning is generated and displayed to the dispatchers. If the task is accepted, the MFMS checks whether the reply is received early enough and, otherwise, warns the dispatchers.

4.4.2.2 Monitoring

The MFMS monitors transportation processes by analysing all incoming messages, determining actual data, comparing actual with planned data, and monitoring whether

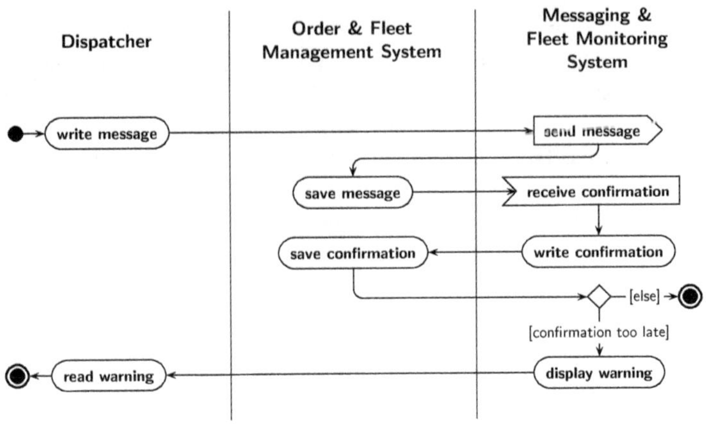

Fig. 4.11: Inform driver

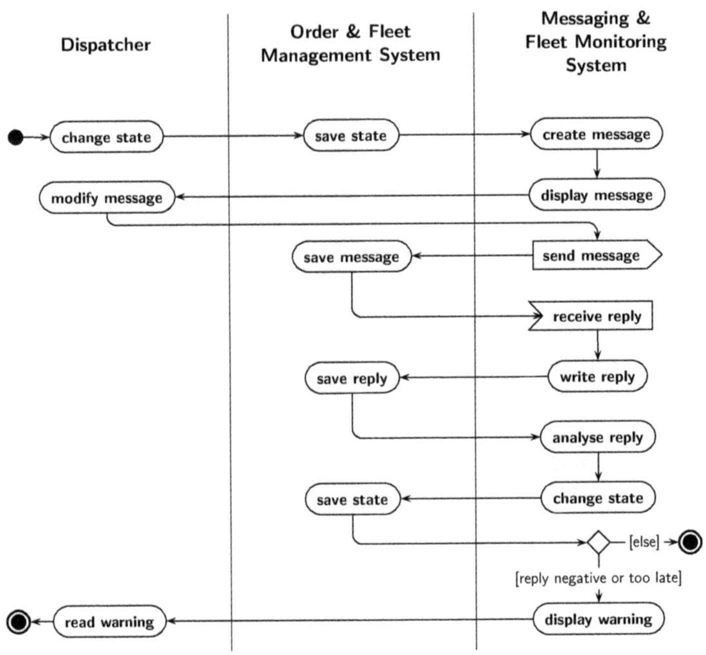

Fig. 4.12: Instruct driver

all planned events occur as expected[1]. The information required to do this has to be obtained from the OFMS, the FCS, and the TTIS and is illustrated in the class diagram in figure 4.13.

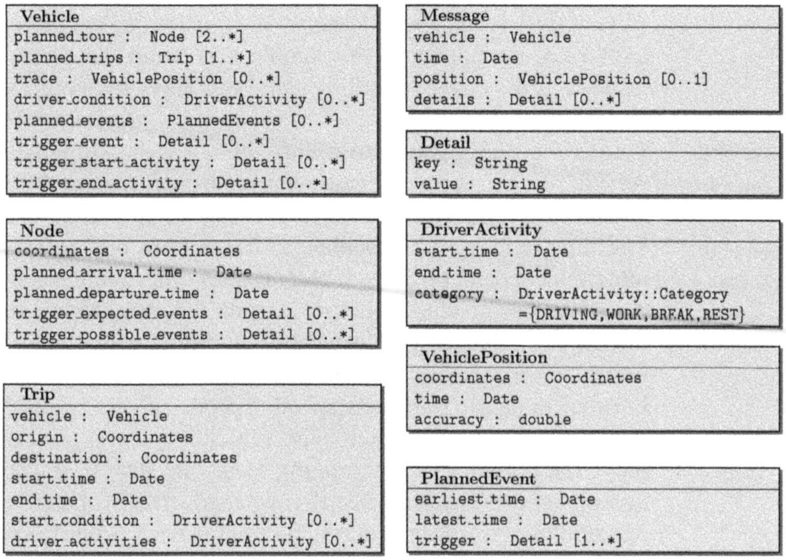

Fig. 4.13: Information required by the MFMS

Detail An instance of this class corresponds to any information detail which may be included in a message. Details are characterised by a key-value pair. This information may be added to the message by driver input or sensor data. An example of details added by driver input is the driver identification (key: "PIN", value: "1234"). An example of details added by sensor data is the opening of the rear door of the cargo body (key: "Rear door", value: "opened").

DriverActivity In order to calculate total travel times, considering regulations regarding drivers' working hours, it is necessary to know what activities a driver has been conducting for how long. Each activity can be categorised as driving, work, break, or rest period.

Message Each incoming message is analysed by the MFMS. Besides of the vehicle and the time, the message may contain the position of the vehicle and a set of details.

[1] Compare with Goel and Gruhn (2006b)

Node An instance of this class corresponds to a point in the tour of a vehicle. A node is represented by the coordinates of the corresponding geographical location and its planned arrival and departure time. Different events can occur at different nodes. Some of them are expected to occur there and some can occasionally occur there. Events are identified if an incoming message contains respective details. An example of a detail indicating such an event is the start of handling activities (key: "Loading", value: "begun").

PlannedEvent Some events are planned to occur within a certain time window. An example of such a planned event is the confirmation that a certain message has been read (key: "Message read", value: "1234").

Trip The time required for a vehicle to travel from one point to another may depend on the vehicle, the start time, and the activities the driver has conducted before starting the trip. During the trip further activities are conducted by the driver, which need to be known in order to calculate the duration of subsequent trips.

Vehicle In order to monitor all transportation processes the planned tour and trips have to be known. The vehicle's trace is used to improve the accuracy of determining the actual state of transportation processes. The activities already conducted by the driver are required to determine the duration of future trips. Some events are planned to occur within a certain time window. Other events are not planned and may occur at any time. An example of such an event is the transmission of odometer information (key: "Odometer", value: "100000"). Furthermore, the start and end of some local activities may be detected if messages contain the corresponding detail information. An example of such a local activity is the start and end of a rest period (key: "Rest period", value: "begun" and key: "Rest period", value: "end").

VehiclePosition If the vehicle is equipped with a positioning system the current vehicle position can be added to messages sent to the dispatching office. Depending on the positioning system the vehicle position may have different accuracy which has to be considered when analysing messages.

If the MFMS is supplied with all necessary data, it can analyse incoming messages, determine actual data, compare it with the planned data, and can monitor whether all planned events occur as expected. For this purpose it detects information which can be represented by the classes illustrated in figure 4.14.

Activity If a vehicle is not moving it can conduct a local activity. The start and end of a local activity are usually identified by the start event and the end event. The duration of the activity can be estimated if the start event is detected. For example, if the begin of han-

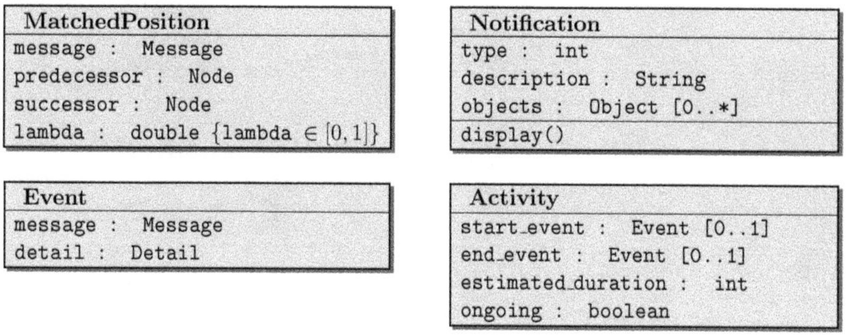

Fig. 4.14: Information detected by the MFMS

dling activities is detected by the details (key: "Loading", value: "begun") the estimated duration can be looked up in the database.

Event An event is a singular occurrence which has triggered the transmission of a specific key-value pair.

MatchedPosition The matched position is the vehicle's position with respect to the planned tour of the vehicle and can be described by two succeeding nodes in the planned tour and a value `lambda` $\in [0, 1]$ indicating the relative position between the nodes.

Notification Any obscurity related to data which cannot be interpreted unambiguously or any detected irregularity result in a notification which is displayed to the dispatchers. The notification may be related to other objects, e.g. a vehicle, a message, etc..

The MFMS has its own database in which all detected information is stored. Furthermore, some of the information may be used to adjust the data in the OFMS. For example, if the database of the OFMS contains a field for the actual arrival time at a handling location, and the vehicle's position is matched to a handling location for the first time, the arrival time at this location can be automatically set by the MFMS.

Message processing

The processing of incoming messages is realised as illustrated in figure 4.15. The MFMS receives an incoming message from the FCS and analyses it in order to determine matched position, events, and activities. All relevant information is adjusted and stored in the OFMS. In exceptional cases, a dispatcher may have modified the tour of a vehicle while the MFMS was analysing the message and the analysis of the message has to be repeated by the MFMS. Notifications are displayed to the dispatchers who can initiate countermeasures if necessary. As a result, dispatchers do not need to read and analyse all incoming messages and thus, can concentrate on those messages which cannot be interpreted automatically or where unexpected incidents are identified.

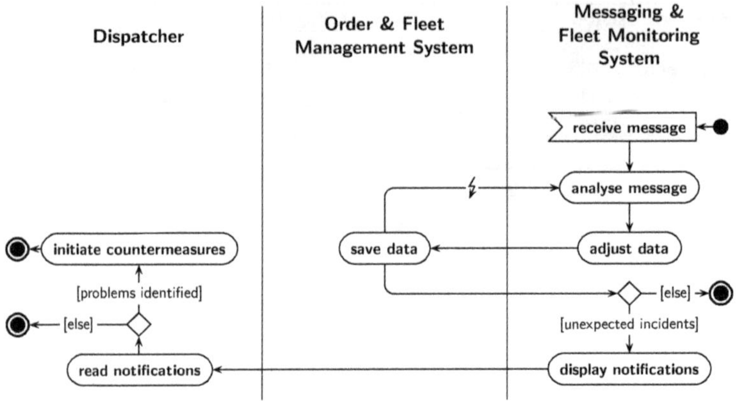

Fig. 4.15: Process message

Analysing messages is a non-trivial task due to the fact that potential discrepancies between planned and actually used route, between planned and actual arrival and departure times, and between expected messages and actually received messages must be considered. Furthermore, inaccuracies of the positioning system, inaccurate planning data, and faulty driver input can increase the complexity of automatic analysis. In figure 4.16 it is illustrated how the MFMS analyses incoming messages.

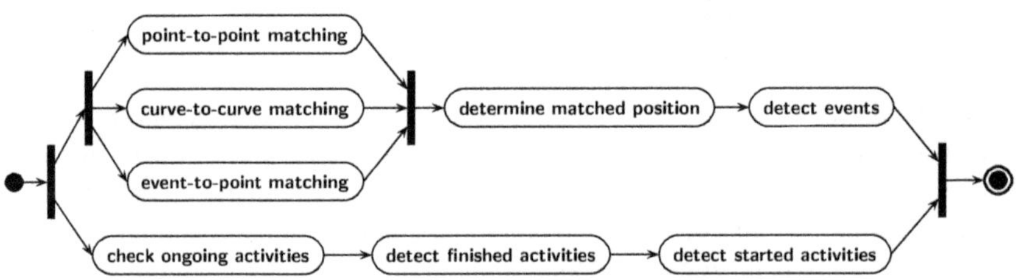

Fig. 4.16: Analyse message

The MFMS determines the vehicle's position with respect to the planned tour. If the message contains a valid position, the coordinates are matched to all nodes in the planned tour which are reasonably "close" regarding planned arrival and departure times. This point-to-point matching is only successful if the vehicle is within vicinity of one of the nodes in the planned tour and if the accuracy of the position is sufficiently good.

Furthermore, the vehicle's trace including the current position is matched to the planned tour. This curve-to-curve matching is similar to the map matching techniques used in in-vehicle navigation systems[1]. In contrast to in-vehicle map matching, however, there is usually less information available to the MFMS. On the other hand, less candidate curves have to be considered. If this curve-to-curve matching is not successful or the distance between the current position and the matched position is too high, the vehicle may have left the planned tour and the dispatchers have to be notified.

If the message contains additional information in form of key-value pairs, these message details are also analysed as certain key-value pairs may indicate events which can only occur at certain nodes in the planned tour. This event-to-point matching has to consider which events are expected at the nodes and which may occasionally occur there.

Obviously, multiple reasonable matches may be found in point-to-point, curve-to-curve, and event-to-point matching. Instead of independently determining the best matched position for each of these three types of matching, a "fitness" value is determined for each type and for each possible match. In the next step, the best matched position is determined by multi-objective optimisation. In the multi-objective optimisation the fitness value of all three types of matchings must be considered. Only those matches can be accepted which are reasonably good for all cases. That is, if a match with a very good fitness value in one of the cases has a bad fitness value in one of the others, this match is unlikely to be a good choice. Dispatchers are notified if no reasonable match can be found. After determining the matched position, the MFMS checks whether events have occurred. If a matched position has been found with `lambda` $=$ `0` or `lambda` $=$ `1`, those events which are possible or expected to occur at the corresponding node can also be detected.

Simultaneously to the processes described above, the MFMS checks all local activities whether they are still ongoing. As local activities are bound to a specific location, an activity cannot be ongoing if the vehicle's position has changed since the start of the activity. In this case, the dispatcher is notified that the activity is no longer ongoing and the end of it has not been detected before. In the next step, message details are analysed to determine whether an activity is finished. If no corresponding ongoing activity can be found, the dispatchers are notified, as an activity is finished which has not been ongoing. Eventually, message details are used to detect activities which just have started.

Event monitoring

Just as an incoming message can give information about possible problems, the non-appearance of messages may indicate that some unexpected incident occurred. In order to store exact arrival and departure times, a message is usually expected at each arrival at and departure from a pickup or delivery location. If such a message is not received by the time it is expected this may indicate some delay. If a driver does not confirm the receipt of an instruction sent to him this may indicate that he has not received or read the instruction. In figure 4.17 the monitoring of such events is illustrated.

[1] See Bernstein and Kornhauser (1996) and White et al. (2000)

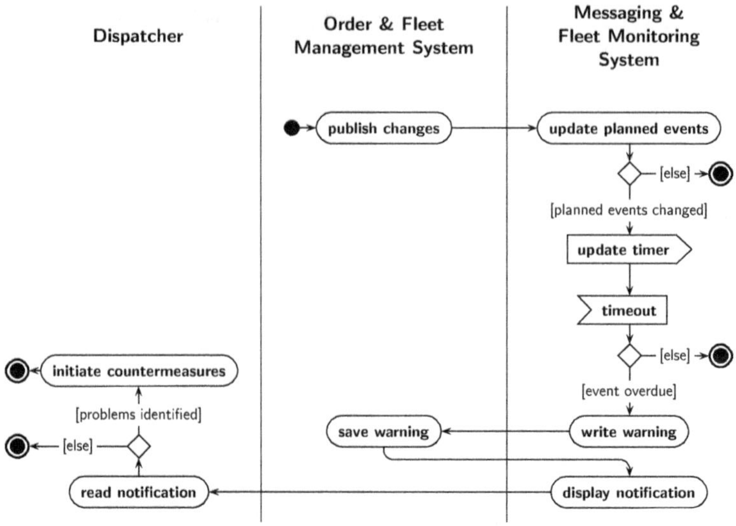

Fig. 4.17: Monitor events

The OFMS publishes all changed data and the MFMS analyses these changes in order to determine whether they are associated to planned events. If the changed data are of no relevance to the planned events, nothing has to be done. Otherwise, the planned events and the corresponding timer are updated. When the updated timer sends a timeout signal, the MFMS checks whether the event is overdue or whether it has occurred before. If an event is overdue, a warning message is generated and stored in the OFMS. Eventually, this warning is shown to the dispatcher who can initiate countermeasures if any problems are identified.

4.4.3 Real-time decision support

The lack of timely and reliable information used to be a major obstacle for computer-based real-time decision support. The telematics-enabled information system alleviates this obstacle as it automatically retrieves actual data from the FTS. Computer-based real-time decision support can be provided by a Dynamic Planning System (DPS) as described in Goel and Gruhn (2006c). Figure 4.18 illustrates how the DPS is integrated into the telematics-enabled information system. All data required for computer-based decision support must be retrieved from the OFMS and the MFMS. In order to provide a unifying interface, the OFMS and the MFMS can be encapsulated within a wrapper package. Thus, the DPS does not have to be developed considering the specific interfaces of the legacy OFMS and the MFMS. In order to view and modify the schedule, the DPS takes over the role of the dispatcher as illustrated in figure 4.19. In the real-time decision support system dispatchers, MFMS,

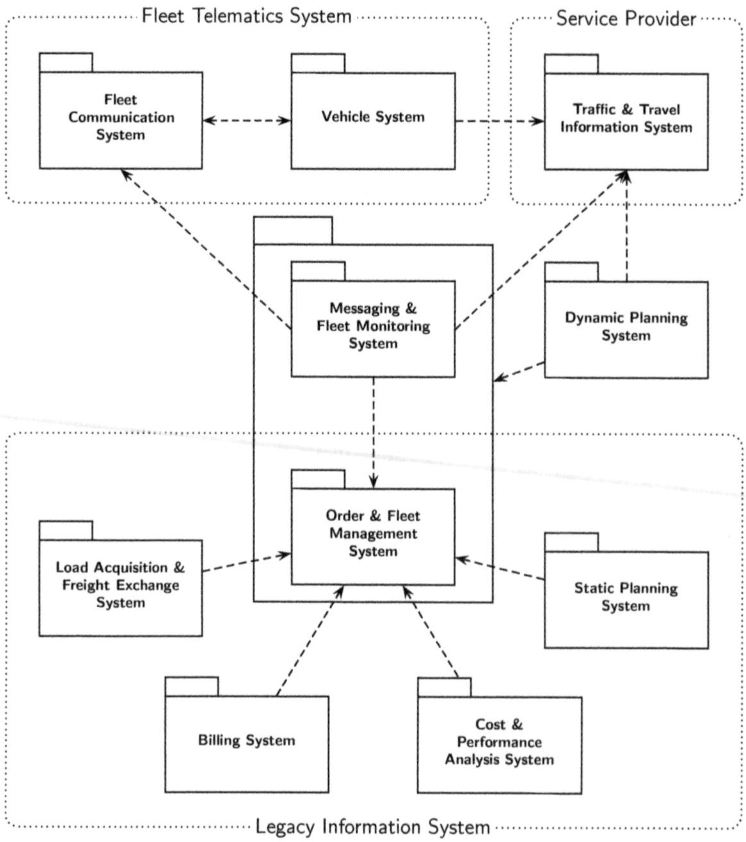

Fig. 4.18: The Dynamic Planning System

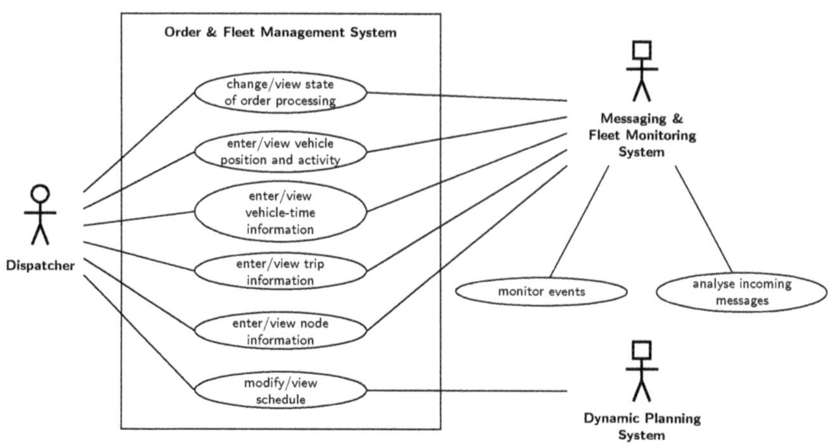

Fig. 4.19: The DPS takes over the role of the dispatchers

and DPS can cooperatively modify problem data and schedule using different problem knowledge and solution methods:

Dispatcher Human dispatchers are usually in direct contact with drivers and shippers and spend a major part of their time to collect and verify problem data[1]. They are able to recognise patterns, can appreciate the consequences if restrictions should be temporarily relaxed and finally can estimate the value of the balance between conflicting objectives[2]. They can use historical problem knowledge and intuition for the creative solving of unknown or new problems. Furthermore, they are very good at challenging the information that is in the computer and augmenting computer-provided solutions with head knowledge.

MFMS As the MFMS analyses all messages sent by the vehicles, it can quickly identify discrepancies between planned and actual data. The MFMS does not modify the schedule itself, however, it revises problem data such that disturbances in the transportation system can be considered by other actors.

DPS The DPS takes the current schedule and uses algorithms to find improving solutions to the analytical model. As computers are very good at processing vast quantities of information, they can quickly determine and evaluate a large number of alternative dispatching decisions.

The real-time decision support system allows these specialised actors to concurrently modify problem data and solution. Therefore, an efficient transaction control scheme must be applied to ensure data consistency. As the DPS continuously tries to optimise the schedule, it must not get an exclusive lock on data records. Otherwise, dispatchers and MFMS would not be able to modify the locked data records. An optimistic locking scheme can be applied in order to prevent that changes made by one actor conflict with changes made by another actor. Optimistic locking does not lock data records when they are read, and proceeds on the assumption that the data records being updated will not be changed[3]. To ensure data consistency, the optimistic locking scheme proposed involves reading a *Transaction Control Number* (TCN) along with each data record representing an order or a tour. When the schedule is changed by the dispatchers or the DPS, the TCNs of the respective order and tour are written back to the database when the record is updated. A pre-update trigger checks the value of the updated TCNs against those held in the database. If the TCNs do not match, the transaction is rejected. With each successful update, the TCNs of the order and tour are incremented. Furthermore, if dispatchers or the MFMS change attributes of data records concerning orders or tours, the TCNs are also incremented. The increment of the TCNs ensures that no other transaction that holds the same TCN will be able to modify the record.

The DPS uses this optimistic locking scheme as illustrated in figure 4.20. First, the DPS gets a snapshot of the current schedule. Due to disturbances in the trans-

[1] See Powell et al. (2002)

[2] See Kopfer and Schönberger (2002)

[3] See Elmasri and Navathe (2000) and Thornton (2001)

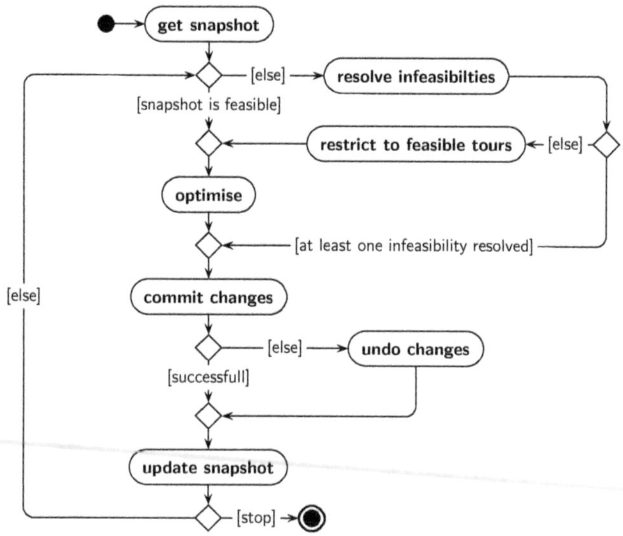

Fig. 4.20: Dynamic planning

portation processes, some tours may not be feasible according to the underlying optimisation model. For example, a delayed vehicle may not be able to reach all pickup and delivery locations within given time windows. In this case, the infeasibility can be resolved by assigning the respective order to another vehicle if the driver of the delayed vehicle has not yet been informed to serve the order, or to remove the order from the tour if it is not yet confirmed[1]. The DPS tries to automatically resolve all infeasibilities and commit respective changes. Suppose that none of the infeasibilities can be resolved automatically. Then, all infeasible tours and all orders assigned to these tours are removed from the optimisation model. Thus, the optimisation method only considers feasible tours, orders assigned to feasible tours, and unscheduled orders. After invoking the optimisation method, the DPS tries to commit all changes made to the schedule. If successful the current snapshot is updated in order to consider all meanwhile changes in the data. Otherwise, changes made on the snapshot by the DPS are undone before the snapshot is updated. The DPS continues with the next iteration if no stopping condition is met - which, of course, never is the case if the system is used in a rolling planning horizon.

The dynamic optimisation of schedules is a nontrivial task. First, the underlying optimisation model must be capable of handling the real-life requirements. Second, many optimisation methods need a lot of time to calculate solutions - even for simple models. The optimisation method used by the DPS, however, must have fast response times. That is, the time needed per iteration must be very fast. Otherwise, it would be very unlikely that the optimised schedule can be applied as dispatchers or the MFMS

[1] See section 3.5.2.2

may have modified relevant data during optimisation. The next chapters will present models and algorithms suitable for real-time decision support.

4.4.4 Supply chain integration

The inter-organisational integration of information systems can benefit due to the better availability of timely and reliable information. However, the relevant subsystems of the LIS usually cannot exploit the full potentials of the improved information supply. These subsystems can either be modernised or replaced in order to provide additional functionalities. As these subsystem usually are not as complex as the OFMS, a replacement of these systems is not as risky and costly. In this section a possible

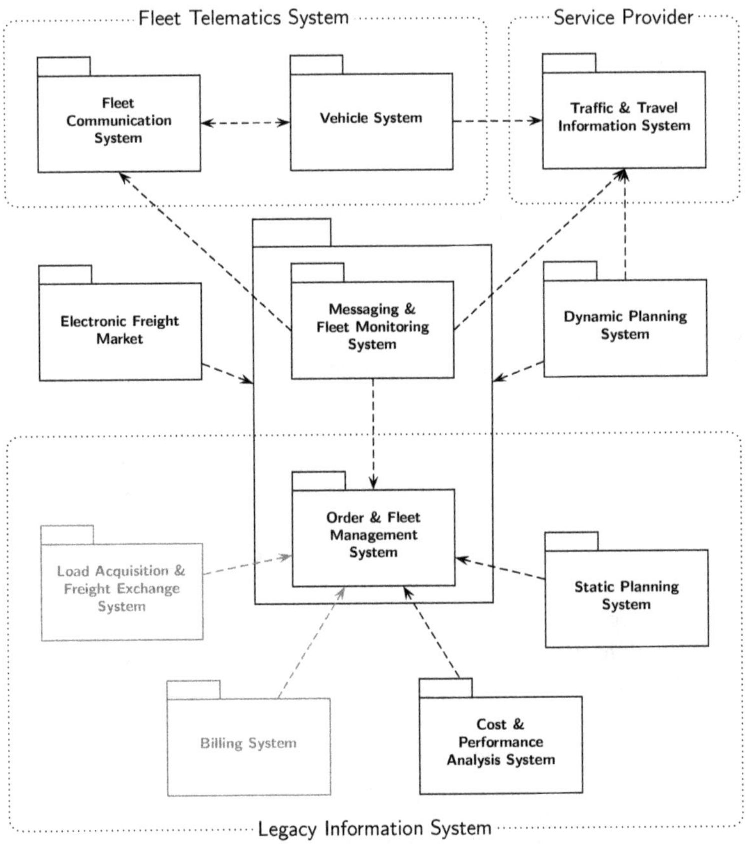

Fig. 4.21: The Electronic Freight Market

replacement of the LAFES and the BS by an Electronic Freight Market (EFM) will be briefly outlined. The EFM may replace the LAFES and the BS or may be deployed as an additional subsystem as illustrated in figure 4.21. The EFM can be used to support the entire life cycle from the acquisition of an order until the payment of the transportation service as illustrated in figure 4.22.

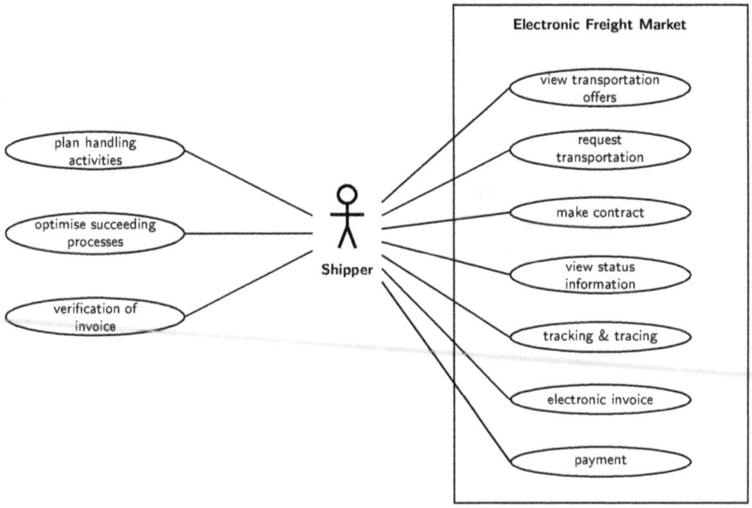

Fig. 4.22: Functions of the Electronic Freight Market

Tracking & tracing functionalities and arrival time estimations can be supplied within the EFM to inform shippers about the current progress of transportation. This information can be used by shippers to plan handling activities and to optimise succeeding tasks. Furthermore, the higher level of information quality in terms of timely availability and reliability also significantly helps in the process of verification of invoices. This speeds up invoicing and can result in earlier payment of transportation services. As information provided to shippers is of much higher quality, there is no need for shippers to get additional information from the dispatchers. This gives further relief for the dispatchers. A more detailed discussion of electronic freight markets is out of scope of this work and the reader is referred to Bierwirth et al. (2002).

4.5 Implementation and case study

This section describes the LIS of the motor carrier introduced in the case study presented in section 3.6. It shows how the LIS was used prior to the deployment of a FTS and the problems the carrier had to face. The FTS which was deployed in order to tackle these problems and the MFMS developed in order to integrate the FTS into the LIS are described. Eventually, a prototype of the DPS is presented which can be integrated into the telematics-enabled information system.

The core of the LIS is a proprietary OFMS especially designed to meet the carrier's business requirements. The OFMS uses an *Oracle 9i Database* to which all OFMS clients connect by using the *Oracle Call Interface* (OCI) as provided with the *Oracle 9i Database Client*[1]. The OFMS clients provide graphical user interfaces (GUI) as shown in figure 4.23. Vehicle tours are visualised in a Gantt-chart manner. The time a transportation request is being fulfilled and the time required for empty replacement trips is represented by a vertical panel. Information about the trips is displayed within the panels, e.g. the names of pickup and delivery locations. The state of order processing is indicated by the colour of the panel. Vehicle type and capacity are indicated by different pictograms in the header of each column. An overview of all currently unscheduled transportation requests is provided by a separate view. The GUI allows dispatchers to modify the schedule and to add additional information and comments.

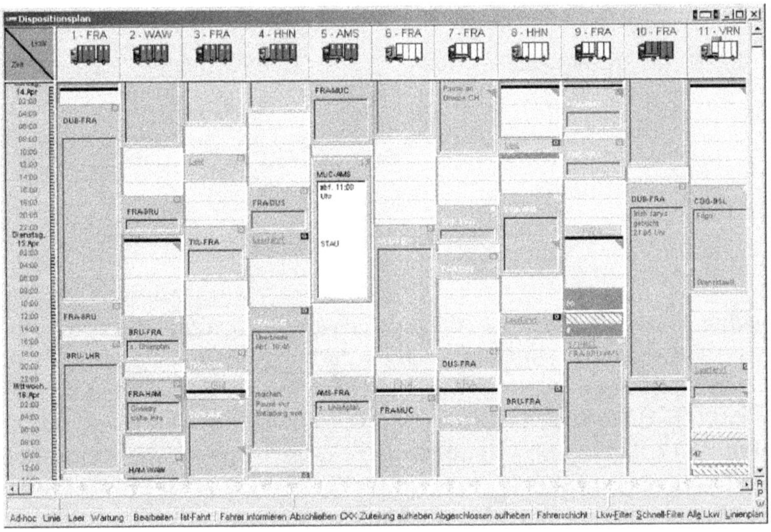

Fig. 4.23: GUI of the OFMS client

In the past the entire communication between drivers and dispatchers was realised by mobile phones. Drivers were instructed to report all particular occurrences to the dispatchers. Dispatchers, however, were often busy talking to shippers, and drivers could not talk to them. Furthermore, background noise in the vehicle and the dispatching office (where several dispatchers may have been telephoning at the same time) increased the risk of misunderstandings. All relevant information had to be manually entered into the OFMS bearing the risk of typing errors. As a result, it was not uncommon that the information in the way-bill did not match the information

[1] See Oracle (2006)

stored in the OFMS. The way-bill was usually handed over to the dispatching office only some time after completion of transportation processes. At this time, however, drivers may have forgotten what really had happened, making it impossible to clarify possible ambiguities.

Another major problem used to be, that dispatchers often had no or only very imprecise and out-of-date information about the actual state of the transportation system. Therefore, dispatchers had no means of identifying discrepancies between actual and planned data. Delays were often identified only when it was too late to initiate countermeasures. Even if a delay of one vehicle was identified early, transportation requests could not be assigned to another vehicle, as the progress of transportation processes of other vehicles was not known. Therefore, important level of service characteristics such as punctuality and flexibility had a potential for improvement.

For these reasons, and because airlines were increasingly requiring that shipments could be located at any time, the motor carrier decided to deploy a fleet telematics system. The vehicles were equipped with mobile *fleetec III* systems which can communicate with the stationary *DATAfleet* system[1]. The *fleetec III* system is shown in figure 4.24 and consists of a display, configurable status buttons, and a GPS receiver. The communication is realised by using the Short Message Service (SMS) provided by the Global System for Mobile Communications (GSM). Figure 4.25 shows a screenshot of the *DATAfleet* system which allows to communicate with drivers, includes a message history, and provides digital maps to visualise vehicle positions.

The *fleetec III* systems are configured to regularly send messages containing the vehicle's position. The frequency, in which these messages are sent, can be adjusted in order to reduce communication costs. Furthermore, the vehicle systems automatically send messages when the vehicle starts or stops moving. To avoid unnecessarily high costs these messages are only sent if certain thresholds are exceeded. In order to be able to detect exact arrival and departure times, which are required for invoicing, drivers are instructed to push predefined status buttons when they arrive at (or depart from) a pickup or delivery location. In order to be able to calculate estimated arrival times, driving and rest periods must be known in the dispatching office. Therefore, drivers are instructed to push status buttons indicating the begin or end of rest periods. The *fleetec III* system automatically adds odometer information and, if available, the vehicle's position to all messages sent.

We implemented a MFMS for Georgi Transporte in the years 2002-2003. The MFMS has been successfully tested and is in operation since 2003. For the most part, the MFMS implemented corresponds with the MFMS described in section 4.4.2. For confidentiality it can only be described with a low level of detail.

The MFMS connects to the *DATAfleet* system using the *DATAfleet ODBC Interface*[2]. The interface allows to access message text, vehicle ID, timestamp, longitude, latitude, and odometer information of all messages received. Messages can be sent

[1] See datafactory AG (2005)

[2] See datafactory AG (2002)

Source: datafactory AG (2005)

Fig. 4.24: *fleetec III*

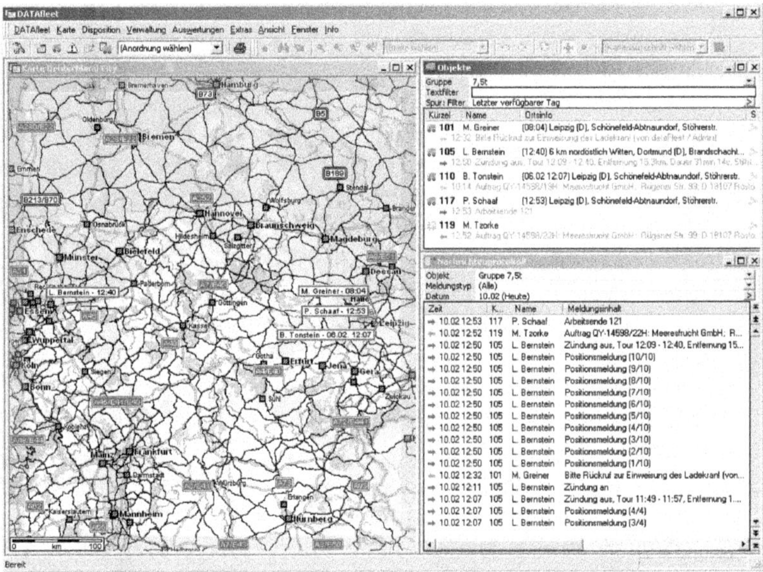

Source: datafactory AG (2005)

Fig. 4.25: *DATAfleet*

by inserting respective data records into a special table of the *DATAfleet* database. Like the OFMS clients, the MFMS connects to the database of the OFMS by using the *Oracle Call Interface* (OCI). Furthermore, the MFMS has its own database in which parameters, planned events, driver activities, details about planned trips, as well as detected activities and events, matched positions, notifications, and arrival time estimates are stored.

Drivers are now instructed about new tasks by text messages instead of telephone calls. A GUI is provided to allow dispatchers to select scheduled orders and compose messages to the drivers. The MFMS automatically creates a message template containing the flight number, the name of pickup and delivery locations, as well as pickup and delivery times. To avoid misunderstandings this message is shown to the dispatcher in the same way it will be displayed on the vehicle system. The dispatchers may edit the message before sending it. After the message is sent, the MFMS automatically adjusts the state of order processing. If the new task is rejected or not confirmed until a certain deadline, the MFMS automatically displays a notification to warn the dispatchers.

In order to analyse all incoming messages, the MFMS first parses the message text to identify key-value pairs. Then, the MFMS analyses all incoming messages as described in section 4.4.2.2. If no matched position is determined, dispatchers are notified that the message cannot be automatically analysed. If the vehicle's position is too far away from its estimated position, the vehicle may be delayed or may have deviated from its planned tour and a notification is shown to the dispatchers.

The arrival at a pickup or delivery location can be detected for two reasons. First, the driver may have pushed a status button to report the arrival. Second, the vehicle's position may be matched to the pickup or delivery location. In both cases, information relevant for invoicing, in particular arrival time and odometer information are automatically adjusted in the OFMS. If no message was sent by the driver, the MFMS notifies the dispatchers that an expected message is not yet received. Whenever the MFMS identifies the arrival at a pickup or delivery location, it automatically calculates the planned departure time by adding the planned handling time to the actual arrival time. If no departure message is sent until this time (plus a small threshold), dispatchers are notified that there may be a delay in handling operations. The cause for this delay has to be determined by the dispatchers for invoicing reasons, e.g. by calling the driver.

All vehicle activities can be determined, as messages are automatically sent whenever the vehicle starts or stops moving, and as drivers are instructed to push status buttons whenever they begin or end a rest period. Analogously to planned departure times at pickup or delivery locations, the MFMS automatically calculates the planned end of a rest period by adding the regular duration of rest periods to the start time. If the vehicle does not continue driving until this time (plus a small threshold), dispatchers are notified that there may be a problem.

As all driving and rest periods are known, arrival time estimates can be calculated considering drivers' working hours. If one of the nodes in the planned tour cannot be reached within given time windows, the MFMS creates a notification to warn the dispatchers.

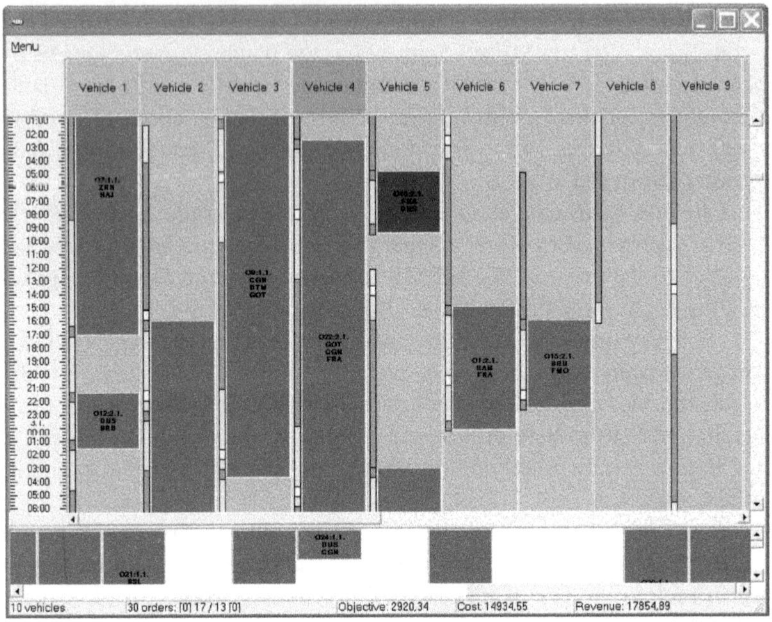

Fig. 4.26: GUI of the prototype

In order to provide a proof of concept, we developed a prototype of the DPS described in section 4.4.3. The prototype was implemented at the Chair of Applied Telematics/e-Business (University of Leipzig) in the year 2005. As it was only developed as a proof of concept it was not integrated into the proprietary OFMS described above. Instead, a simple GUI was implemented which allows to visualise the current schedule. The GUI allows to manually modify the current schedule by dragging panels corresponding to transportation requests to the tour of a vehicle. The prototype of the DPS allows interactive optimisation, that is, it allows human user and optimisation method to collaboratively optimise the current schedule. Simultaneous changes made by the human user and the DPS are accepted if they concern different tours and orders. However, when concurrent changes by the human user and the DPS involve the same orders and tours, the changes of either the human user or the DPS are rejected. In order to avoid that dispatching decisions made by the human user are immediately undone by the DPS, all manual decisions are automatically fixed. Furthermore, the human user may manually fix and unfix any part of the schedule at any time. The DPS restricts the set of tours which may be modified to those tours which are not fixed. Furthermore, scheduled orders which are fixed within a tour are treated as route restrictions similar to those described on page 43. Figure 4.26 shows a screenshot of the GUI giving an overview of the current schedule and drivers' working hours including driving times, handling times, breaks, and daily rest periods. In the figure, the tour of *Vehicle 4* and the order assigned to the tour of *Vehicle 5* are manually fixed by the human user and will not be changed by the DPS.

The prototype developed does not need to consider that tours may be infeasible, as human users are only allowed to do feasible changes. However, analogously to tours fixed by the human user, infeasible tours could be removed from the optimisation model. The optimisation method and the underlying optimisation model are presented in the next chapters.

5

Models for routing a fleet of commercial vehicles

5.1 Introduction

In order to provide computer-based decision support, the real-life problem a motor carrier has to face must be represented by an *analytical model*, i.e. an abstraction of the real world which allows the automatic determination and evaluation of schedules. This chapter presents an overview of models for routing a fleet of commercial vehicles. The history of vehicle routing models dates back to 1959 when Dantzig and Ramser formulated the *Truck Dispatching Problem*[1]. They described an application concerning the delivery of gasoline between a bulk terminal and several service stations supplied by the terminal. Since then, hundreds of models and thousands of algorithms have been proposed in the vehicle routing literature. These problems have in common the determination of the optimal set of tours to be performed by a fleet of vehicles $\mathcal{V}$ to serve a set of transportation requests $\mathcal{O}$. They generalise the well-known *Travelling Salesman Problem* (TSP) which is the problem of finding the least cost tour (for one vehicle) to visit a finite number of customers. The TSP is extensively described in Lawler et al. (1985) and has been introduced under the name *Botenproblem* (Messenger Problem) by Karl Menger in 1930 at a mathematical colloquium in Vienna[2]. The TSP is known to be $\mathcal{NP}$-complete even if arc costs satisfy the triangle inequality[3]. Thus, all of the problems discussed in this chapter are also $\mathcal{NP}$-complete as they generalise the TSP.

In road transport the movements of vehicles are restricted to the road network, the topology of which can be obtained from Geographical Information Systems[4]. As the original road network is too complex to be considered directly, the original road network is usually transformed into a reduced network $(\mathcal{N}, \mathcal{A})$. The nodes of the reduced network represent customer locations or the depot. In order to distinguish different customers at the same geographical location, the set of nodes may include

[1] See Dantzig and Ramser (1959)

[2] See Menger (1932)

[3] See Ahuja et al. (1993)

[4] See section 2.2.3

different nodes referring to the same geographical location. For each pair of nodes $n, m \in \mathcal{N}$ the arc $(n, m) \in \mathcal{A}$ is associated to the shortest path starting at node n and ending at node m. Due to the presence of one-way streets and other regulations, the shortest path in the original road network usually depends on the direction of traversing a road and the type of vehicle. The set of arcs $\mathcal{A} \subseteq \mathcal{N} \times \mathcal{N} \setminus \{(n, n) \mid n \in \mathcal{N}\}$ includes at most one arc from one node to another. Networks with *multi-arcs*[1], i.e. several arcs in the same direction with different costs and travel times, can usually be considered with little additional effort.

In this work the term *route* is used for any path in the original road network representing the roads a vehicle is travelling on. The term *tour* is used to represent a sequence of nodes in the network $(\mathcal{N}, \mathcal{A})$ visited by a vehicle. Each tour is associated with a route which can be derived by replacing arcs between two successive nodes by the corresponding shortest paths in the original road network. Thus, finding the least cost route to cover a set of customer locations is equivalent to finding the least cost tour.

This chapter surveys several classical vehicle routing models and presents mathematical formulations of the models. It begins with a formulation of the Vehicle Routing Problem (VRP) which is the simplest and most studied model for routing a fleet of commercial vehicles. Then other classical models, namely the Vehicle Routing Problem with Time Windows (VRPTW), the Heterogeneous Fleet Vehicle Routing Problem with Time Windows (HFVRPTW), and the Pickup and Delivery Problem with Time Windows (PDPTW) are described.

Real-life vehicle routing problems encounter a variety of practical complexities which, to a certain extend, have been considered by the classical models. However, the classical models often oversimplify the problems that occur in practice. It appears, that in the vehicle routing literature more effort has been made in finding good solutions for the classical models, as for developing models that can handle the requirements occurring in real-life problems[2]. However, rich models, that can handle the requirements arising in real-life problems, are a fundamental prerequisite for sophisticated computer-based decision support. A model that can handle most of of the real-life requirements, can significantly reduce the effort to manually verify an automatically generated schedule and the effort to transform the model recommendation into an applicable schedule. Hence, it can be concluded that the development of good models is equally important as the development of good algorithms for solving the problems.

This chapter introduces a general model that can handle the complexities evolving from various characteristics found in real-life vehicle routing problems that are not considered by the classical models. The model will be termed the *General Vehicle Routing Problem* (GVRP) and unifies the formulations of the classical vehicle routing problems as illustrated in figure 5.1.

Like the classical models, the GVRP assumes that travel times are constant, usually proportional to the distance travelled. In many real-life applications, however,

[1] See Ahuja et al. (1993)

[2] See Kilby et al. (2000)

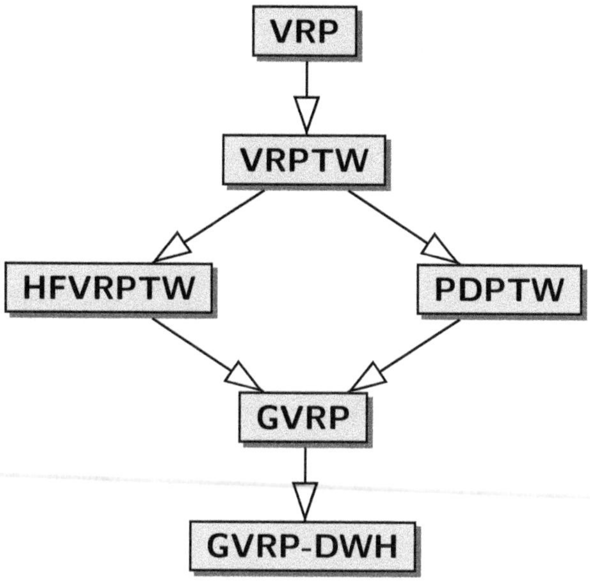

Fig. 5.1: Vehicle routing models and their relationships

regulations concerning drivers' working hours must be considered. Such regulations are of particular importance as they have significant impact on total travel times, i.e. the time required for driving, breaks, and rest periods. This chapter shows how drivers' working hours can be considered in vehicle routing models and concludes by introducing the *General Vehicle Routing Problem with Drivers' Working Hours* (GVRP-DWH).

5.2 The Vehicle Routing Problem

The *Vehicle Routing Problem* (VRP) concerns the distribution of goods and products between a depot and customers. Real-life applications of the VRP are found in the delivery or collection of shipments. The VRP was first presented by Dantzig and Ramser (1959) who describe the problem of delivering gasoline from a bulk terminal to service stations. Other delivery problems arise in various industries, for example, food and beverages, newspaper, and postal carriers. Collection problems occur e.g. in manufacturing when parts and components have to be transported to the production plant, in the collection of fresh milk, and the collection of garbage. This section presents formulations of the VRP and some of its variants. Further information about the VRP, its variants, and applications can be found in the book on the VRP edited by Toth and Vigo (2002).

Let us consider a set of customers who require the delivery or the collection of a certain shipment. Let $\mathcal{C}$ denote the set of customer locations and let n_{depot} denote the depot where all vehicles start and end their tour. Let

$$\mathcal{N} := \mathcal{C} \cup \{n_{\text{depot}}\}$$

and

$$\mathcal{A} := \mathcal{N} \times \mathcal{N} \setminus \{(n,n) \mid n \in \mathcal{C}\}.$$

Each arc $(n,m) \in \mathcal{A}$ is associated with a nonnegative cost c_{nm} which usually represents the required travel time or the travelled distance on the arc. Each customer location $n \in \mathcal{C}$ is associated with a known resource demand or supply r_n. It is assumed that either all customers $n \in \mathcal{C}$ require a certain amount of goods ($r_n < 0$) or request the pickup of a certain amount of goods ($r_n > 0$). The capacity of the vehicles is denoted by $r^{\max}$ and $r_{n_{\text{depot}}} \in [0, r^{\max}]$ denotes the initial load of the vehicle at the depot. Usually, $r_{n_{\text{depot}}} = r^{\max}$ in delivery problems and $r_{n_{\text{depot}}} = 0$ in collection problems.

Definition 1 *A sequence of nodes $\theta = (n_1, \ldots, n_\lambda)$ is a VRP tour if and only if*

- $n_1 = n_{depot}$ *and* $n_\lambda = n_{depot}$
- $n_i \in \mathcal{C}$ *for all* $1 < i < \lambda$
- *for all* $i, j \in \{2, \ldots, \lambda - 1\}$ *: if* $n_i = n_j$ *then* $i = j$.

Definition 2 *A VRP tour $\theta = (n_1, \ldots, n_\lambda)$ is feasible according to capacity constraints*[1] *if and only if*

$$0 \leq \sum_{j=1}^{j \leq i} r_{n_j} \leq r^{max} \textit{ for all } 1 < i < \lambda.$$

The VRP is the problem of finding feasible tours covering all customers, such that each customer is visited exactly once, and that the cost for operating the tours is minimised. The VRP can be modelled using the two-index binary variables x_{nm} indicating whether $m \in \mathcal{N}$ is visited immediately after node $n \in \mathcal{N}$ by some vehicle ($x_{nm} = 1$), or not ($x_{nm} = 0$). For each node $n \in \mathcal{N}$ let ρ_n denote a variable representing the current load at the node.

The Vehicle Routing Problem (VRP) is

minimise

$$\sum_{(n,m) \in \mathcal{A}} x_{nm} c_{nm} \tag{I.1}$$

[1] Usually only a single resource attribute is considered by the models found in the literature, e.g. weight. However, multiple resource attributes can easily be considered in the capacity constraints by representing the supplied and demanded resources by vectors. In this case all operations and comparisons have to be understood element wise.

subject to

$$\sum_{(n,m)\in\mathcal{A}} x_{nm} = \sum_{(m,n)\in\mathcal{A}} x_{mn} \text{ for all } n \in \mathcal{N} \tag{I.2}$$

$$\sum_{(n,m)\in\mathcal{A}} x_{nm} = 1 \text{ for all } n \in \mathcal{C} \tag{I.3a}$$

$$\sum_{\substack{(n,m)\in\mathcal{A}\\ n=n_{\text{depot}}}} x_{nm} = |\mathcal{V}| \tag{I.3b}$$

$$\rho_{n_{\text{depot}}} = r_{n_{\text{depot}}} \tag{I.4a}$$

$$\text{for all } (n,m) \in \mathcal{A} \text{ with } m \in \mathcal{C}: \text{ if } x_{nm} = 1 \text{ then } \rho_m = \rho_n + r_m \tag{I.4b}$$

$$0 \leq \rho_n \leq r^{\max} \text{ for all } n \in \mathcal{N} \tag{I.4c}$$

$$x_{nm} \in \{0,1\} \text{ for all } (n,m) \in \mathcal{A} \tag{I.5}$$

The objective function (I.1) represents the accumulated costs for all arcs used in the solution. Equation (I.2) represents the flow conservation constraints which impose that exactly the same number of vehicles reach a node $n \in \mathcal{N}$ as vehicles depart from the same node. Equation (I.3a) imposes that each customer is exactly visited once and equation (I.3b) imposes that all vehicles leave the depot. Note that $(n_{\text{depot}}, n_{\text{depot}}) \in \mathcal{A}$ and thus, vehicles do not have to visit any customer location. Constraints (I.4a), (I.4b), and (I.4c) are the capacity constraints which impose that the accumulated load is within the capacity of the vehicle at each node. Eventually, constraint (I.5) imposes that all values of x_{nm} are binary.

This two-index formulation is the most basic formulation of the VRP and many variants of the VRP have been proposed in the literature to consider additional real-life requirements.

5.2.1 Time window restrictions

The *Vehicle Routing Problem with Time Windows* (VRPTW) is a generalisation of the VRP in which each customer $n \in \mathcal{C}$ is associated with a time interval $[t_n^{\min}, t_n^{\max}]$, called a *time window*. All customers have to be reached within the specified time window. A vehicle may have to wait on the trip towards a node such that it arrives within the time window of the node. Let $t_{n_{\text{depot}}}$ denote a parameter representing the earliest departure time at the depot. For each customer location $n \in \mathcal{C}$ let t_n denote a variable representing the arrival time at the customer. In order to consider the time required to travel from node $n \in \mathcal{N}$ to node $m \in \mathcal{N}$, the parameter d_{nm} representing the travel time is specified. The service time needed at a customer location may be included in the travel times of arcs leaving the customer. Throughout this work it is assumed that travel times are positive and obey the *triangle inequality*

$$d_{ij} + d_{jk} \geq d_{ik} \text{ for all } i, j, k \in \mathcal{N}.$$

The triangle inequality expresses that it is never faster to travel between any two nodes by visiting an intermediate node as to directly travel from one node to another. This assumption has little affect on generality as this property usually is satisfied in real-life applications.

Definition 3 *A VRP tour $\theta = (n_1, \ldots, n_\lambda)$ is feasible according to time windows if and only if arrival times $t_{n_2}, \ldots, t_{n_{\lambda-1}}$ exist such that*

$$t_{n_i} + d_{n_i n_{i+1}} \leq t_{n_{i+1}} \text{ for all } 1 \leq i < \lambda - 1$$

and

$$t^{min}_{n_i} \leq t_{n_i} \leq t^{max}_{n_i} \text{ for all } 1 < i < \lambda.$$

The arrival times of a VRP tour $\theta = (n_1, \ldots, n_\lambda)$ can be calculated in $O(\lambda)$ time by

$$t_{n_{i+1}} := \max\left(t^{\min}_{n_{i+1}}, t_{n_i} + d_{n_i n_{i+1}}\right)$$

for all $1 \leq i < \lambda - 1$. These arrival times are feasible if $t_{n_i} \leq t^{\max}_{n_i}$ for all $1 < i < \lambda$. Otherwise, the tour θ is infeasible according to time windows.

The Vehicle Routing Problem with Time Windows (VRPTW) is

minimise

$$\sum_{(n,m)\in\mathcal{A}} x_{nm} c_{nm} \tag{II.1}$$

subject to

$$\sum_{(n,m)\in\mathcal{A}} x_{nm} = \sum_{(m,n)\in\mathcal{A}} x_{mn} \text{ for all } n \in \mathcal{N} \tag{II.2}$$

$$\sum_{(n,m)\in\mathcal{A}} x_{nm} = 1 \text{ for all } n \in \mathcal{C} \tag{II.3a}$$

$$\sum_{\substack{(n,m)\in\mathcal{A} \\ n=n_{\text{depot}}}} x_{nm} = |\mathcal{V}| \tag{II.3b}$$

$$\rho_{n_{\text{depot}}} = r_{n_{\text{depot}}} \tag{II.4a}$$

$$\text{for all } (n,m) \in \mathcal{A} \text{ with } m \in \mathcal{C}: \text{ if } x_{nm} = 1 \text{ then } \rho_m = \rho_n + r_m \tag{II.4b}$$

$$0 \leq \rho_n \leq r^{\max} \text{ for all } n \in \mathcal{N} \tag{II.4c}$$

$$\text{for all } (n,m) \in \mathcal{A} \text{ with } m \in \mathcal{C}: \text{ if } x_{nm} = 1 \text{ then } t_m \geq t_n + d_{nm} \tag{II.5a}$$

$$t_n^{\min} \leq t_n \leq t_n^{\max} \text{ for all } n \in \mathcal{C} \tag{II.5b}$$

$$x_{nm} \in \{0,1\} \text{ for all } (n,m) \in \mathcal{A} \tag{II.6}$$

In this formulation (II.1) to (II.4c) and (II.6) are identical to (I.1) to (I.4c) and (I.5). Constraints (II.5a) and (II.5b) are the time window constraints. Constraint (II.5a) imposes that each node can only be reached according to the arrival time of the preceding node and the (positive) travel time between the two nodes. Inequality (II.5b) imposes that each arrival time is within the time window of the customer.

5.2.2 Heterogeneous vehicle fleet

The *Heterogeneous Fleet Vehicle Routing Problem with Time Windows* (HFVRPTW) is a generalisation in which the vehicle fleet is composed of different vehicles with different properties. Vehicles may have different travel costs, travel times, and capacity, and may be located at different depots. The two-index formulation does not give any information about which vehicle traverses an arc. Thus, the two-index formulation cannot be used for the HFVRPTW. One way to overcome this drawback is to explicitly indicate the vehicle by using the three-index binary variable x^v_{nm} indicating whether vehicle $v \in \mathcal{V}$ travels on arc $(n,m) \in \mathcal{A}$ ($x^v_{nm} = 1$) or not ($x^v_{nm} = 0$).

Assume that each vehicle $v \in \mathcal{V}$ is located at depot n_v and let $\mathcal{D} := \{n_v \mid v \in \mathcal{V}\}$ denote the set of depots. Without loss of generality it is assumed that $n_v \neq n_w$ for any two vehicles $v, w \in \mathcal{V}$[1]. The network underlying the problem formulation is defined by

$$\mathcal{N} := \mathcal{C} \cup \mathcal{D}$$

and

$$\mathcal{A} := \mathcal{N} \times \mathcal{N} \setminus \{(n,n) \mid n \in \mathcal{C}\}.$$

Let c^v_{nm} and d^v_{nm} denote costs and travel times of vehicle $v \in \mathcal{V}$ travelling on arc $(n,m) \in \mathcal{A}$. The earliest departure time of vehicle $v \in \mathcal{V}$ at its depot n_v is denoted by t_{n_v}. The capacity of vehicle $v \in \mathcal{V}$ is denoted by r^v and the initial load of the vehicle is denoted by r_{n_v}.

Definition 4 *A sequence of nodes $\theta = (n_1, \ldots, n_\lambda)$ is a HFVRPTW tour of a vehicle $v \in \mathcal{V}$ if and only if*

- $n_1 = n_v$ *and* $n_\lambda = n_v$
- $n_i \in \mathcal{C}$ *for all* $1 < i < \lambda$
- *for all* $i, j \in \{2, \ldots, \lambda - 1\}$: *if* $n_i = n_j$ *then* $i = j$

[1] Note that two different nodes may be associated to the same geographical location.

Definition 5 *A HFVRPTW tour $\theta = (n_1, \ldots, n_\lambda)$ of vehicle $v \in \mathcal{V}$ is feasible according to capacity constraints if and only if*

$$0 \leq \sum_{j=1}^{j \leq i} r_{n_j} \leq r^v \text{ for all } 1 < i < \lambda.$$

Definition 6 *A HFVRPTW tour $\theta = (n_1, \ldots, n_\lambda)$ of vehicle $v \in \mathcal{V}$ is feasible according to time windows if and only if arrival times $t_{n_2}, \ldots, t_{n_{\lambda-1}}$ exist such that*

$$t_{n_i} + d^v_{n_i n_{i+1}} \leq t_{n_{i+1}} \text{ for all } 1 \leq i < \lambda - 1$$

and

$$t^{min}_{n_i} \leq t_{n_i} \leq t^{max}_{n_i} \text{ for all } 1 < i < \lambda.$$

The Heterogeneous Fleet Vehicle Routing Problem with Time Windows (HF-VRPTW) is
minimise

$$\sum_{v \in \mathcal{V}} \sum_{(n,m) \in \mathcal{A}} x^v_{nm} c^v_{nm} \tag{III.1}$$

subject to

$$\sum_{(n,m) \in \mathcal{A}} x^v_{nm} = \sum_{(m,n) \in \mathcal{A}} x^v_{mn} \text{ for all } v \in \mathcal{V}, n \in \mathcal{N} \tag{III.2}$$

$$\sum_{v \in \mathcal{V}} \sum_{(n,m) \in \mathcal{A}} x^v_{nm} = 1 \text{ for all } n \in \mathcal{N} \tag{III.3a}$$

$$\sum_{\substack{(n,m) \in \mathcal{A} \\ n = n_v}} x^v_{nm} = 1 \text{ for all } v \in \mathcal{V} \tag{III.3b}$$

$$\rho_{n_v} = r_{n_v} \text{ for all } v \in \mathcal{V} \tag{III.4a}$$

$$\text{for all } v \in \mathcal{V}, (n,m) \in \mathcal{A} \text{ with } m \in \mathcal{C}: \text{ if } x^v_{nm} = 1 \text{ then } \rho_m = \rho_n + r_m \tag{III.4b}$$

$$\text{for all } v \in \mathcal{V}, n \in \mathcal{C}: \text{ if } \sum_{(n,m) \in \mathcal{A}} x^v_{nm} = 1 \text{ then } 0 \leq \rho_n \leq r^v \tag{III.4c}$$

$$\text{for all } v \in \mathcal{V}, (n,m) \in \mathcal{A} \text{ with } m \in \mathcal{C}: \text{ if } x^v_{nm} = 1 \text{ then } t_m \geq t_n + d^v_{nm} \tag{III.5a}$$

$$t^{\min}_n \leq t_n \leq t^{\max}_n \text{ for all } n \in \mathcal{C} \tag{III.5b}$$

$$x^v_{nm} \in \{0, 1\} \text{ for all } (n,m) \in \mathcal{A}, v \in \mathcal{V} \tag{III.6}$$

The objective function (III.1) represents the accumulated costs for all arcs used by some vehicle in the solution. Equation (III.2) represents the flow conservation constraints which impose that each vehicle that reaches a node also departs from it. Equation (III.3a) imposes that each customer is exactly visited once, and equation (III.3b) imposes that each vehicle $v \in \mathcal{V}$ starts its tour at its depot n_v. Constraints (III.4a) to (III.4c) represent the capacity constraints and constraints (III.5a) to (III.5b) represent the time window constraints. Eventually, constraints (III.6) impose that all values of x^v_{nm} are binary.

5.3 The Pickup and Delivery Problem

In the *Pickup and Delivery Problem* (PDP) the set of customer locations $\mathcal{C}$ can be partitioned into a set of pickup locations $\mathcal{C}_P$ and a set of delivery locations $\mathcal{C}_D$. Each transportation request is specified by a load to be transported from its pickup location to its delivery location. The PDP is the problem of finding a set of optimal tours for a fleet of vehicles, in order to serve these transportation requests. In other words, the PDP deals with the construction of optimal tours in order to visit all pickup and delivery locations and to satisfy precedence and pairing constraints. Precedence constraints deal with the restriction that each pickup location has to be visited prior to visiting the corresponding delivery location. Pairing constraints restrict the set of admissible tours such that pickup and delivery locations of each transportation request are visited by the same vehicle. Comprehensive surveys on the PDP are provided by Mitrović-Minić (1998) and Desaulniers et al. (2002). The PDP is also known as *Dial-A-Ride Problem* (DARP) in case of transport of persons. This case, however, is not in the scope of this work and the reader is referred to the survey by Cordeau and Laporte (2003) for more information on the DARP.

The *Pickup and Delivery Problem with Time Windows* (PDPTW) generalises the VRPTW. In the distribution (collection) of goods the VRPTW is the case where the pickup (delivery) locations of transportation requests are all at the common depot. The two-index formulation of the VRPTW, however, cannot be used to model the PDPTW as the knowledge, which vehicle visits a node, is required to guarantee that pairing constraints are satisfied. Instead, the PDPTW is modelled using the three-index binary variable $x_n m^v$ which is also used for formulating the HFVRPTW.

Let $\mathcal{D} := \{n_v \mid v \in \mathcal{V}\}$ denote the set of depots, whereas $n_v \in \mathcal{D}$ denotes the depot of vehicle $v \in \mathcal{V}$. For each transportation request $o \in \mathcal{O}$ let $n_{(o,1)} \in \mathcal{C}_P$ denote the pickup location and let $n_{(o,2)} \in \mathcal{C}_D$ denote the corresponding delivery location. The set of customer locations is $\mathcal{C} := \mathcal{C}_P \cup \mathcal{C}_D$. The network underlying the problem formulation is defined by

$$\mathcal{N} := \mathcal{C} \cup \mathcal{D}$$

and

$$\mathcal{A} := \mathcal{N} \times \mathcal{N} \setminus \{(n,n) \mid n \in \mathcal{C}\}.$$

Let c^v_{nm} and d^v_{nm} denote costs and (positive) travel times of vehicle $v \in \mathcal{V}$ travelling on arc $(n,m) \in \mathcal{A}$. The earliest departure time of vehicle $v \in \mathcal{V}$ at its depot n_v is denoted by t_{n_v}. The capacity of vehicle $v \in \mathcal{V}$ is denoted by r^v and the initial load of the vehicle is denoted by r_{n_v}.

Definition 7 *A sequence of nodes* $\theta = (n_1, \ldots, n_\lambda)$ *is a PDPTW tour of a vehicle* $v \in \mathcal{V}$ *if and only if*

- $n_1 = n_v$ *and* $n_\lambda = n_v$
- $n_i \in \mathcal{C}$ *for all* $1 < i < \lambda$
- *for all* $i, j \in \{2, \ldots, \lambda - 1\}$ *: if* $n_i = n_j$ *then* $i = j$
- *for all* $o \in \mathcal{O}$ *:* $\{n_1, \ldots, n_\lambda\} \cap \{n_{(o,1)}, n_{(o,2)}\} = \emptyset$ *or* $\{n_{(o,1)}, n_{(o,2)}\} \subset \{n_1, \ldots, n_\lambda\}$
- *for all* $o \in \mathcal{O}$, $i, j \in \{2, \ldots, \lambda - 1\}$ *: if* $n_i = n_{(o,1)}$ *and* $n_j = n_{(o,2)}$ *then* $i < j$.

The same definitions of feasibility apply for a PDPTW tour as for a HFVRPTW tour.

The Pickup and Delivery Problem with Time Windows (PDPTW) is

minimise

$$\sum_{v \in \mathcal{V}} \sum_{(n,m) \in \mathcal{A}} x_{nm}^v c_{nm}^v \tag{IV.1}$$

subject to

$$\sum_{(n,m) \in \mathcal{A}} x_{nm}^v = \sum_{(m,n) \in \mathcal{A}} x_{mn}^v \text{ for all } v \in \mathcal{V}, n \in \mathcal{N} \tag{IV.2}$$

$$\sum_{v \in \mathcal{V}} \sum_{(n,m) \in \mathcal{A}} x_{nm}^v = 1 \text{ for all } n \in \mathcal{N} \tag{IV.3a}$$

$$\sum_{\substack{(n,m) \in \mathcal{A} \\ n = n_v}} x_{nm}^v = 1 \text{ for all } v \in \mathcal{V} \tag{IV.3b}$$

$$\rho_{n_v} = r_{n_v} \text{ for all } v \in \mathcal{V} \tag{IV.4a}$$

$$\text{for all } v \in \mathcal{V}, (n,m) \in \mathcal{A} \text{ with } m \in \mathcal{C} : \text{ if } x_{nm}^v = 1 \text{ then } \rho_m = \rho_n + r_m \tag{IV.4b}$$

$$\text{for all } v \in \mathcal{V}, n \in \mathcal{C} : \text{ if } \sum_{(n,m) \in \mathcal{A}} x_{nm}^v = 1 \text{ then } 0 \le \rho_n \le r^v \tag{IV.4c}$$

$$\text{for all } v \in \mathcal{V}, (n,m) \in \mathcal{A} \text{ with } m \in \mathcal{C} : \text{ if } x_{nm}^v = 1 \text{ then } t_m \ge t_n + d_{nm}^v \tag{IV.5a}$$

$$t_n^{\min} \le t_n \le t_n^{\max} \text{ for all } n \in \mathcal{C} \tag{IV.5b}$$

$$t_{n_{(o,1)}} \le t_{n_{(o,2)}} \text{ for all } o \in \mathcal{O} \tag{IV.6}$$

$$\sum_{\substack{(n,m) \in \mathcal{A} \\ n = n_{(o,1)}}} x_{nm}^v = \sum_{\substack{(n,m) \in \mathcal{A} \\ n = n_{(o,2)}}} x_{nm}^v \text{ for all } o \in \mathcal{O}, v \in \mathcal{V} \tag{IV.7}$$

$$x_{nm}^v \in \{0,1\} \text{ for all } (n,m) \in \mathcal{A}, v \in \mathcal{V} \tag{IV.8}$$

In this formulation (IV.1) to (IV.5b) and (IV.8) are identical to (III.1) to (III.5b) and (III.6). Constraint (IV.6) represents the precedence constraints which impose that all pickup locations are visited before the corresponding delivery location. Constraint (IV.7) represents the pairing constraints which impose that pickup and delivery location associated to one order are visited by the same vehicle.

5.4 The General Vehicle Routing Problem

Many real-world applications encounter practical complexities not considered by the classical models. In this section a general model is introduced, which is capable of handling real-life requirements not considered in the VRP, the PDP, and its most popular variants. First, this section discusses some of the problem characteristics which can be found in real-life problems. Then, the General Vehicle Routing Problem (GVRP) originally introduced by Goel and Gruhn (2007) is verbally described and a mathematical formulation of the GVRP is given.

5.4.1 Load acceptance and employment of external carriers

In the classical models it is assumed that load acceptance decisions are made before planning starts. Therefore, the fleet size may have to be increased in order to guarantee that all transportation requests can be served within time windows. In dynamic problems load acceptance and routing decisions are done simultaneously, e.g. when new transportation requests arrive. The flexibility to increase the fleet size is usually very low in dynamic problems and sometimes not all transportation requests can be served by self-operated vehicles. If a transportation requests $o \in \mathcal{O}$ can be served, load acceptance decisions are usually based on the associated revenue p_o. Depending on the specific situation, p_o represents the price the shipper is willing to pay, the penalty costs in case a confirmed transportation request is not served, or the price an external carrier demands for performing the transport. A transportation request is only accepted if the revenue gained is higher than the incremental costs for serving the request.

Among the few works found in the vehicle routing literature on combined load acceptance and vehicle routing problems are the survey on the Travelling Salesman Problem with Profits by Feillet et al. (2005), and the study on the Pickup and Delivery Selection Problem by Schönberger et al. (2002).

5.4.2 Route restrictions

The concept of a depot as start and end point of a tour is often not very useful in dynamic problems[1]. As the borders between load acceptance, routing, and transportation are vanishing, the current position of a vehicle has to be considered each

[1] See Savelsbergh and Sol (1995)

time the schedule is modified. Thus, the depot where the vehicle started its tour is of no interest when diverting a vehicle from its current route.

In dynamic vehicle routing problems some transportation requests may already be partially served at the time of planning. Therefore, it must be guaranteed that all shipments that are already picked up will be transported to the corresponding delivery locations. The current position and load impose route restrictions which must be considered in a model suitable for dynamic vehicle routing.

5.4.3 Arbitrary number of pickup, delivery, and service locations

In the classical models all transportation requests are expected to have the same number of pickup, delivery, and service locations. In the VRP all transportation requests are associated with a single pickup or delivery location. In the PDP all transportation requests concern the pickup of a shipment at its origin and the delivery at its destination. In real-life applications transportation may be more complex and may involve several pickup, delivery, and/or service locations which must be visited in a specified sequence. At each of these locations some shipment(s) with several describing attributes can be loaded or unloaded.

The consideration of multiple pickup, delivery, and/or service locations has found little interest in the vehicle routing literature. A rich vehicle routing problem with multiple pickup and delivery locations has been studied by Savelsbergh and Sol (1995) who presented the General Pickup and Delivery Problem (GPDP). In the GPDP a transportation request is composed of several shipments with different pickup and delivery locations. Each pickup location has to be visited before any delivery location. However, the sequence in which the pickup (or delivery) locations must be visited is not specified.

5.4.4 Problem formulation

In the *General Vehicle Routing Problem* (GVRP) a transportation request is specified by a nonempty set of pickup, delivery and/or service locations which have to be visited in a particular sequence by the same vehicle, the time windows in which these locations have to be visited, and the revenue gained by serving the transportation request. Furthermore, some characteristics can be specified which constrain the possibility of assigning the transportation requests to certain vehicles due to compatibility and capacity constraints. At each location some shipment(s) with several describing attributes can be loaded or unloaded. In contrast to many other commonly known routing problems, not all transportation requests have to be assigned to a vehicle. Instead, a so-called *make-or-buy* decision is necessary to determine whether a transportation request should be assigned to a self-operated vehicle (make) or not (buy).

A fleet of heterogeneous vehicles is available to serve the transportation requests. The vehicles can have different capacities, as well as different travel times and travel costs. The vehicles can transport shipments which require some of the capacity the vehicle supplies. Instead of assuming that each vehicle becomes available at a central depot, each vehicle is given a start location, where it becomes available at a specific time and with a specific load. Furthermore, the vehicles do not have to return to a central depot. Instead, a final location is specified for each vehicle, which has to be reached within a specific time and with a specific load. Each vehicle may have to visit some locations in a particular sequence between leaving its start and reaching its final location. All locations have to be visited within specified time windows. A vehicle may have to wait on the trip towards a location in order to arrive within the corresponding time window.

A tour of a vehicle is a journey starting at the vehicles start location and ending at its final location, passing all other locations the vehicle must visit in the correct sequence, and passing all locations belonging to each transportation request assigned to the vehicle in the correct respective sequence. A tour is *feasible* if and only if compatibility constraints are satisfied for all orders assigned to the tour, and time window and capacity restrictions are satisfied at each point in the tour. The objective is to find distinct feasible tours maximising the profit, which is determined by the accumulated revenue of all served transportation requests, reduced by the accumulated costs for operating these tours.

5.4.5 Mathematical formulation

For all orders $o \in \mathcal{O}$ let $n_{(o,1)}, \ldots, n_{(o,\lambda_o)}$ denote the nodes belonging to the order and let

$$\mathcal{C} := \bigcup_{o \in \mathcal{O}} \{n_{(o,1)}, \ldots, n_{(o,\lambda_o)}\}.$$

For all vehicles $v \in \mathcal{V}$ let $n_{(v,1)}, \ldots, n_{(v,\lambda_v)}$ denote the nodes which must be visited by the vehicle and let

$$\mathcal{D} := \bigcup_{v \in \mathcal{V}} \{n_{(v,1)}, \ldots, n_{(v,\lambda_v)}\}.$$

Let

$$\mathcal{N} := \mathcal{C} \cup \mathcal{D}$$

and

$$\mathcal{A} := \mathcal{N} \times \mathcal{N} \setminus \{(n,n) \mid n \in \mathcal{N}\}.$$

For each node $n \in \mathcal{N}$ lower and upper bounds specifying the time windows are denoted by $t_n^{\min}$ and $t_n^{\max}$. For each vehicle $v \in \mathcal{V}$ the (positive) travel time for an arc $(n,m) \in \mathcal{A}$ including some possible service time at node n is denoted by d_{nm}^v. For each vehicle $v \in \mathcal{V}$ the cost for travelling from node $n \in \mathcal{N}$ to node $m \in \mathcal{N}$ is denoted by c_{nm}^v. For each order $o \in \mathcal{O}$ the revenue gained when the order is served is denoted by p_o. Let δ_{ov} denote whether order $o \in \mathcal{O}$ may be served by vehicle $v \in \mathcal{V}$ ($\delta_{ov} = 1$), or not ($\delta_{ov} = 0$). Every vehicle supplies some (typically multi-dimensional) non-negative resource r^v (the capacity). At every node some shipments, which require or release a certain amount of the resource the vehicle supplies, may be loaded or unloaded. For every $n \in \mathcal{N}$ let r_n denote the (typically multi-dimensional) amount of resource requirements for the shipments loaded or unloaded at the node. All operations and comparisons concerning resource requirements and supply have to be understood element wise. If a shipment is loaded r_n is non-negative, if it is unloaded r_n is non-positive.

Definition 8 *A sequence of distinct nodes $\theta = (n_1, \dots, n_\lambda)$ is a GVRP tour of a vehicle $v \in \mathcal{V}$ if and only if*

- $n_1 = n_{(v,1)}$ *and* $n_\lambda = n_{(v,\lambda_v)}$
- *a subset $\mathcal{O}_\theta \subseteq \mathcal{O}$ exists with*

$$\{n_1, \dots, n_\lambda\} = \{n_{(v,1)}, \dots, n_{(v,\lambda_v)}\} \cup \bigcup_{o \in \mathcal{O}_\theta} \{n_{(o,1)}, \dots, n_{(o,\lambda_o)}\}$$

- *for all $i, j \in \{1, \dots, \lambda\}$, $k, l \in \{1, \dots, \lambda_v\}$:*

$$\text{if } n_i = n_{(v,k)} \text{ and } n_j = n_{(v,l)} \text{ and } k < l \text{ then } i < j.$$

- *for all $o \in \mathcal{O}$, $i, j \in \{1, \dots, \lambda\}$, $k, l \in \{1, \dots, \lambda_o\}$:*

$$\text{if } n_i = n_{(o,k)} \text{ and } n_j = n_{(o,l)} \text{ and } k < l \text{ then } i < j.$$

Definition 9 *A GVRP tour $\theta = (n_1, \dots, n_\lambda)$ of vehicle $v \in \mathcal{V}$ is feasible according to compatibility constraints if and only if*

$$\delta_{ov} = 1 \text{ for all } o \in \mathcal{O}_\theta.$$

Definition 10 *A GVRP tour $\theta = (n_1, \dots, n_\lambda)$ of vehicle $v \in \mathcal{V}$ is feasible according to capacity constraints if and only if*

$$0 \leq \sum_{j=1}^{j \leq i} r_{n_j} \leq r^v \text{ for all } 1 \leq i \leq \lambda.$$

Definition 11 *A GVRP tour $\theta = (n_1, \dots, n_\lambda)$ of vehicle $v \in \mathcal{V}$ is feasible according to time windows if and only if arrival times $t_{n_1}, \dots, t_{n_\lambda}$ exist such that*

$$t_{n_i} + d^v_{n_i n_{i+1}} \leq t_{n_{i+1}} \text{ for all } 1 \leq i < \lambda$$

and

$$t^{min}_{n_i} \leq t_{n_i} \leq t^{max}_{n_i} \text{ for all } 1 \leq i \leq \lambda.$$

The GVRP is the problem of finding distinct feasible tours, maximising the profit determined by the accumulated revenue of all orders served by a self-operated vehicle, reduced by the cost for operating the tours.

The GVRP can be modelled using the three-index binary variable x^v_{nm} indicating whether $m \in \mathcal{N}$ is visited immediately after node $n \in \mathcal{N}$ by vehicle $v \in \mathcal{V}$ ($x^v_{nm} = 1$), or not ($x^v_{nm} = 0$). For notational reasons the binary variables y^v_n are added to the formulation. y^v_n indicates whether node $n \in \mathcal{N}$ is visited by vehicle $v \in \mathcal{V}$ ($y^v_n = 1$), or not ($y^v_n = 0$). For each node $n \in \mathcal{N}$ the GVRP contains the variables t_n and ρ_n. If node $n \in \mathcal{N}$ is visited by a vehicle, t_n specifies the arrival time and ρ_n specifies the current load of the vehicle. If no vehicle visits this node, t_n and ρ_n are without any meaning.

The contribution of each vehicle $v \in \mathcal{V}$ to the objective function is

$$\sum_{o \in \mathcal{O}} y^v_{n_{(o,1)}} p_o - \sum_{(n,m) \in \mathcal{A}} x^v_{nm} c^v_{nm}.$$

The first term represents the accumulated revenue of served orders, the second term represents the accumulated costs for vehicle movements.

The General Vehicle Routing Problem (GVRP) is

maximise

$$\sum_{v \in \mathcal{V}} \Big(\sum_{o \in \mathcal{O}} y^v_{n_{(o,1)}} p_o - \sum_{(n,m) \in \mathcal{A}} x^v_{nm} c^v_{nm} \Big) \tag{V.1}$$

subject to

$$\sum_{(n,m) \in \mathcal{A}} x^v_{nm} = \sum_{(m,n) \in \mathcal{A}} x^v_{mn} \text{ for all } v \in \mathcal{V}, n \in \mathcal{N} \tag{V.2}$$

$$y^v_n = \sum_{(n,m) \in \mathcal{A}} x^v_{nm} \text{ for all } v \in \mathcal{V}, n \in \mathcal{N} \tag{V.3a}$$

$$\sum_{v \in \mathcal{V}} y^v_n \leq 1 \text{ for all } n \in \mathcal{N} \tag{V.3b}$$

$$\rho_{n_{(v,1)}} = r_{n_{(v,1)}} \text{ for all } v \in \mathcal{V} \tag{V.4a}$$

$$\text{for all } v \in \mathcal{V}, (n,m) \in \mathcal{A} \text{ with } m \neq n_{(v,1)} : \text{ if } x^v_{nm} = 1 \text{ then } \rho_m = \rho_n + r_m \tag{V.4b}$$

$$\text{for all } v \in \mathcal{V}, n \in \mathcal{N} : \text{ if } y^v_n = 1 \text{ then } 0 \leq \rho_n \leq r^v \tag{V.4c}$$

$$\text{for all } v \in \mathcal{V}, (n,m) \in \mathcal{A} \text{ with } m \neq n_{(v,1)} : \text{ if } x^v_{nm} = 1 \text{ then } t_n + d^v_{nm} \leq t_m \tag{V.5a}$$

$$t^{\min}_n \leq t_n \leq t^{\max}_n \text{ for all } n \in \mathcal{N} \tag{V.5b}$$

$$t_{n_{(v,\mu)}} \leq t_{n_{(v,\mu+1)}} \text{ for all } v \in \mathcal{V}, 1 \leq \mu < \lambda_v \tag{V.6a}$$

$$t_{n_{(o,\mu)}} \leq t_{n_{(o,\mu+1)}} \text{ for all } o \in \mathcal{O}, 1 \leq \mu < \lambda_o \tag{V.6b}$$

$$y^v_{n_{(v,\mu)}} = 1 \text{ for all } v \in \mathcal{V}, 1 \leq \mu \leq \lambda_v \tag{V.7a}$$

$$\sum_{\mu=1}^{\mu \leq \lambda_o} y^v_{n_{(o,\mu)}} = \lambda_o y^v_{n_{(o,1)}} \text{ for all } o \in \mathcal{O}, v \in \mathcal{V} \tag{V.7b}$$

$$y^v_{n_{(o,1)}} \leq \delta_{ov} \text{ for all } o \in \mathcal{O}, v \in \mathcal{V} \tag{V.8}$$

$$x^v_{nm} \in \{0,1\} \text{ for all } v \in \mathcal{V}, (n,m) \in \mathcal{A}, y^v_n \in \{0,1\} \text{ for all } v \in \mathcal{V}, n \in \mathcal{N} \tag{V.9}$$

The objective function (V.1) represents the accumulated revenue of all served orders reduced by the costs for all arcs used in the solution. Equation (V.2) represents the flow conservation constraints which impose that each vehicle reaching a node $n \in \mathcal{N}$ also departs from the node. Constraints (V.3a) and (V.3b) impose that each node is visited at most once. Constraints (V.4a) to (V.4c), and (V.5a) and (V.5b) represent capacity and time window constraints. (V.6a) and (V.6b) are the precedence constraints imposed on the sequence in which nodes associated to vehicles and orders are visited. Equation (V.7a) imposes that each vehicle visits all nodes associated to it. Equation (V.7b) represents the grouping constraint which imposes that all locations belonging to an order are visited by the same vehicle. Inequality (V.8) represents the compatibility constraints which impose that orders are only assigned to vehicles capable of serving the order. Eventually, constraints (V.9) impose that the values of x^v_{nm} and y^v_n are binary.

The GVRP is a generalisation of the classical models described in the previous sections. In contrast to the classical models, not all transportation requests must be served in the GVRP. The requirements that all transportation requests are served, however, can be fulfilled by assigning a sufficiently large revenue p_o to each transportation request $o \in \mathcal{O}$. This guarantees that any solution in which all transportation requests are served has a higher objective function value than every solution in which at least one transportation request is not served.

5.5 Drivers' working hours

The consideration of drivers' working hours in vehicle routing and scheduling is of extraordinary importance as the corresponding regulations usually have a big impact on total travel times, i.e. the time required for driving, breaks, and rest periods. Although of particular importance for many real-life applications, restrictions to drivers' working hours have only received very little attention in the vehicle routing literature. A maximum number of working hours during a tour can be modelled similar to capacity restrictions. This approach is for example discussed by Campbell and Savelsbergh (2004). Only very few works have tried to address vehicle routing problems in which obligatory breaks and rest periods must be scheduled within a tour which may last several working days. Savelsbergh and Sol (1998) study a dynamic and generalised Pickup and Delivery Problem in which lunch breaks and night breaks must be taken within fixed time intervals. Drivers' working hours as regulated by the U.S. Department of Transportation have been considered by Xu et al. (2003) who study a rich Pickup and Delivery Problem. Xu et al. (2003) do not consider that daily rest periods may be taken before the maximum daily driving time is exhausted. Such "early" rest periods, which are required in order to be able to satisfy narrow time windows at subsequent customer locations, are considered by Goel and Gruhn (2006a) who introduce the *Vehicle Routing Problem with Drivers' Working Hours* (VRP-DWH) in which breaks and daily rest periods as regulated by EU social legislation are considered. This section introduces the General Vehicle Routing Problem with Drivers' Working Hours (GVRP-DWH) which generalises the VRP-DWH and considers the following regulations[1]:

- After a driving period of four and a half hours a driver shall take an uninterrupted break of not less than 45 minutes, unless he takes a rest period. During a break a driver must not drive or undertake any other work.
- The daily driving time between the end of one daily rest period and the beginning of the following daily rest period shall not exceed 9 hours. A daily rest period is any period of at least 11 hours during which a driver may freely dispose of his time.
- In a multiple manned vehicle, the other driver(s) may take a break on the moving vehicle whilst one driver is driving.
- In a multiple manned vehicle, the daily rest period in which the vehicle must be stationary may be reduced to 9 hours.
- The weekly driving time shall not exceed 56 hours.
- A weekly rest period shall start no later than 144 hours after the end of the previous weekly rest period.

The last of these regulations can be enforced by imposing respective time window constraints on the start and end node of the vehicle's tour. In this section it is assumed that service times are working periods in which drivers perform handling activities. Consequently, the service time must not be interpreted as a break and must not be part of a daily rest period. Furthermore, it is assumed that only the working periods between two consecutive weekly rest periods have to be considered.

In order to model the GVRP-DWH the following parameters are required:

s_n^v the service time required at node $n \in \mathcal{N}$

δ_{nm}^v the pure driving time from node $n \in \mathcal{N}$ to node $m \in \mathcal{N}$

[1] Note that not all of the regulations described in section 3.3.2.3 are considered.

t^v_{weekly} the maximum weekly driving time between two consecutive weekly rest periods

t^v_{daily} the maximum daily driving time between two consecutive daily rest periods

t^v_{nonstop} the maximum nonstop driving time between two consecutive breaks or rest periods

t^v_{rest} the time required for a daily rest period

t^v_{break} the time required for a break

For all nodes $n \in \mathcal{N}$ a label

$$l_n = \begin{pmatrix} l_{n,1} \\ l_{n,2} \\ l_{n,3} \\ l_{n,4} \end{pmatrix} = \begin{pmatrix} \text{arrival time} \\ \text{weekly driving time} \\ \text{daily driving time} \\ \text{nonstop driving time} \end{pmatrix}$$

can be used to represent the state of the driver at the node. The vehicle can start the service at node $n \in \mathcal{N}$ at time $l_{n,1}$ and can depart from n at time $l_{n,1} + s^v_n$. It may drive $t^v_{\text{weekly}} - l_{n,2}$ before the next weekly rest period, $t^v_{\text{daily}} - l_{n,3}$ before the next daily rest period, and $t^v_{\text{nonstop}} - l_{n,4}$ before the next break.

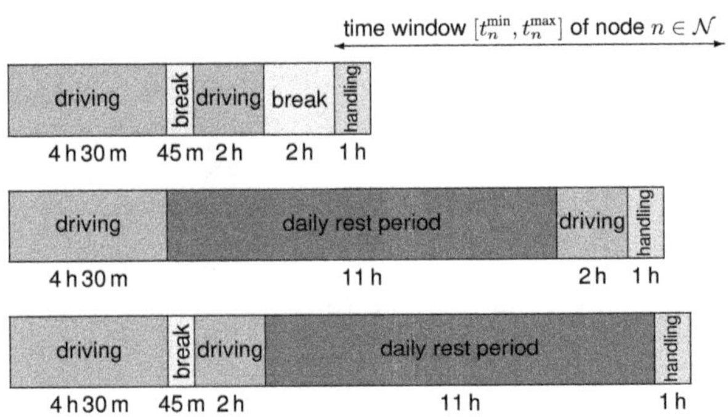

Fig. 5.2: Alternative driver states at a node

As illustrated in figure 5.2, the state of the driver at a node cannot always be uniquely determined, as it is possible to schedule driving and rest periods in a way that breaks and rest periods are taken before the respective accumulated driving time is exhausted.

Although, there may be very many different labels at a node, not all of them need to be considered. A label l_n *dominates* another label l'_n if

$$l_{n,1} \leq l'_{n,1} \text{ and } l_{n,2} \leq l'_{n,2} \text{ and } l_{n,3} \leq l'_{n,3} \text{ and } l_{n,4} \leq l'_{n,4} \tag{D1}$$

or

$$l_{n,1} + t^v_{\text{break}} \leq l'_{n,1} \text{ and } l_{n,2} \leq l'_{n,2} \text{ and } l_{n,3} \leq l'_{n,3} \tag{D2}$$

or

$$l_{n,1} + t^v_{\text{rest}} \leq l'_{n,1} \text{ and } l_{n,2} \leq l'_{n,2}. \tag{D3}$$

Obviously, if a label is dominated by (D1) it does not need to be considered. If a label dominates another label by (D2) the vehicle could continue its tour with a break period, and after the break (D1) would be satisfied. Analogously, if a label dominates another label by (D3) the vehicle could continue its tour with a daily rest period, and after the daily rest period (D1) would be satisfied. Note that none of the labels corresponding to the driver states illustrated in figure 5.2 dominates the others.

Consider that a vehicle is supposed to travel from node $n \in \mathcal{N}$ with label l_n to a node $m \in \mathcal{N}$. As illustrated in figure 5.2, several labels can be determined for node m. As only working periods between two consecutive weekly rest periods are considered, it is assumed that $l_{n,2} + \delta^v_{nm} \leq t^v_{\text{weekly}}$. Otherwise, a weekly rest period would be required before reaching node m. Let $\mathscr{L} := \emptyset$ and let $l_m := (l_{n,1} + s^v_n, l_{n,2} + \delta^v_{nm}, l_{n,3}, l_{n,4})^T$. In order to determine possible labels at node m the recursive function illustrated in figure 5.3 is invoked by `expand_label(`l_m, δ^v_{nm}`)`.

First, the time the vehicle may drive continuously is calculated and the label and the remaining required driving time δ are respectively adjusted. If $\delta = 0$ node m is reached and the label l_m is added to the set $\mathscr{L}$. If $\delta > 0$ and $l_{m,3} = t^v_{\text{daily}}$ a daily rest period is required before the vehicle may continue to travel towards m and the label is respectively adjusted. If $\delta > 0$ and $l_{m,3} < t^v_{\text{daily}}$ and if δ or the remaining daily driving time are less or equal t^v_{nonstop} a new label l'_m is generated with arrival time $l_{m,1} + t^v_{\text{rest}}$, zero daily driving time, and zero nonstop driving time. This new label is generated as it may be beneficial to continue with a daily rest period instead of a break. The new label is expanded by invoking `expand_label(`l'_m, δ`)`. Then, the time required for a break is added to the arrival time of label l_m and the remaining required nonstop driving time is respectively adjusted. The label expansion continues with the calculation of the next driving period.

Let $l_m \in \mathscr{L}$ be a label generated as described above. The arrival time $l_{m,1}$ may be smaller than the begin $t^{\min}_m$ of the time window. Therefore, for all $l_m \in \mathscr{L}$ let

$$\begin{aligned}\mathscr{L}(l_m) := \Big\{ & \big(\max\{t^{\min}_m, l_{m,1}\}, l_{m,2}, l_{m,3}, l_{m,4}\big)^T, \\ & \big(\max\{t^{\min}_m, l_{m,1} + t^v_{\text{break}}\}, l_{m,2}, l_{m,3}, 0\big)^T, \\ & \big(\max\{t^{\min}_m, l_{m,1} + t^v_{\text{rest}}\}, l_{m,2}, 0, 0\big)^T \Big\}\end{aligned}$$

denote a set of potential labels. Now, let $\mathscr{L}^v_m(l_n)$ denote the set of labels containing all

$$l \in \{l'_m \in \bigcup_{l_m \in \mathscr{L}} \mathscr{L}(l_m) \mid l'_{m,1} \leq t^{\max}_m\}$$

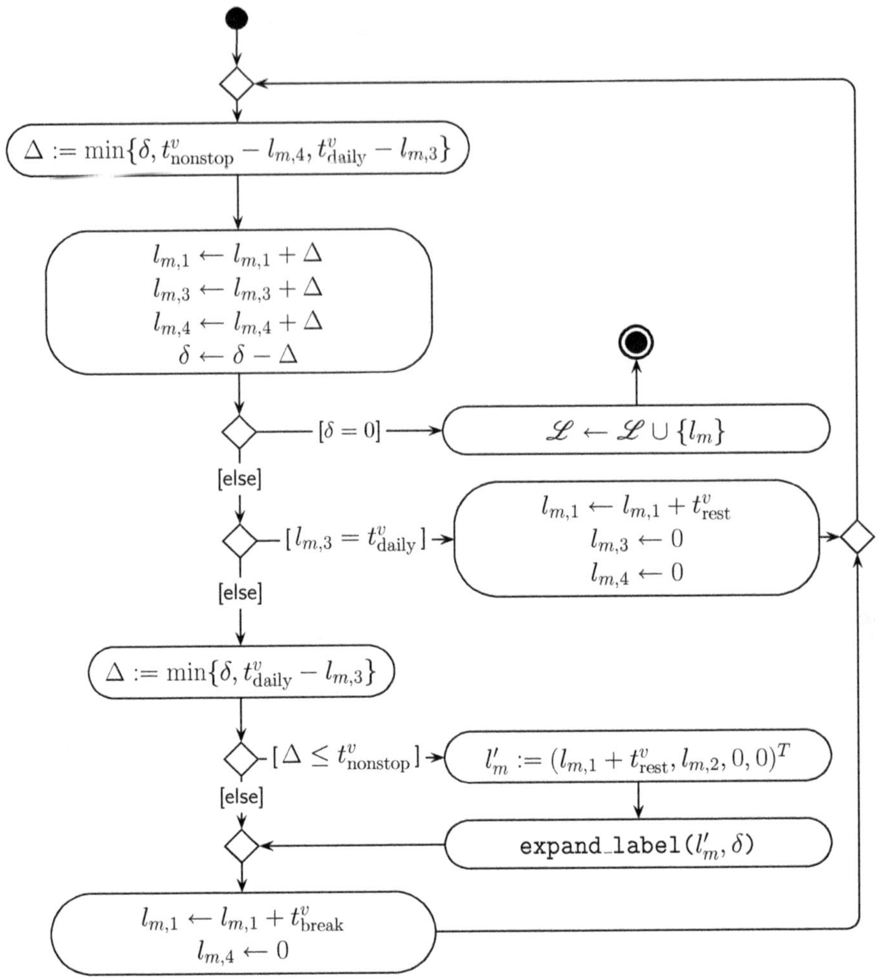

Fig. 5.3: Recursive function `expand_label` (l_m, δ)

such that no label is dominated by any other label and let l^v denote the initial label at the start of the tour of vehicle $v \in \mathcal{V}$.

Definition 12 *A GVRP tour $\theta = (n_1, \dots, n_\lambda)$ of vehicle $v \in \mathcal{V}$ is feasible according to drivers' working hours if and only if labels $l_{n_1}, \dots, l_{n_\lambda}$ exist such that*

$$l_{n_1} = l^v$$

and

$$l_{n_{i+1}} \in \mathscr{L}^v_{n_{i+1}}(l_{n_i}) \text{ for all } 1 \leq i < \lambda.$$

The GVRP with Drivers' Working Hours (GVRP-DWH) is

minimise

$$(V.1)$$

subject to

$$(V.2), (V.3a), (V.3b), (V.4a), (V.4b), (V.4c),$$

and

$$l_{n_{(v,1)}} = l^v \text{ for all } v \in \mathcal{V} \tag{VI.5a}$$

$$\text{for all } (n,m) \in \mathcal{A}, v \in \mathcal{V} \text{ with } m \neq n_{(v,1)}: \text{ if } x^v_{nm} = 1 \text{ then } l_m \in \mathscr{L}^v_m(l_n) \tag{VI.5b}$$

$$l_{n,1} = t_n \text{ for all } n \in \mathcal{N} \tag{VI.5c}$$

and

$$(V.6a), (V.6b), (V.7a), (V.7b), (V.8), (V.9)$$

Constraints (VI.5a) to (VI.5c) replace the time window constraints (V.5a) and (V.5b) and impose that drivers' working hours are satisfied for all trips. The Pickup and Delivery Problem with Drivers' Working Hours and the Heterogeneous Fleet Vehicle Routing Problem with Drivers' Working Hours can be modelled analogously. Note that constraint (VI.5c) is only required to ensure that precedence constraints are satisfied and thus, is not required for the Heterogeneous Fleet Vehicle Routing Problem with Drivers' Working Hours.

5.6 Case study

The routing problem of the motor carrier introduced in the case study in section 3.6 cannot be modelled by the classical models due to the variety of practical complexities. However, it can be modelled as a GVRP-DWH as shown in this section.

For simplification only a small problem with two vehicles and three orders is considered. Both vehicles are based in Frankfurt where they start and end their tour. Vehicle v_1 has a capacity of 5D, i.e. five ULD of type D. The cargo body has tautliner sides and the vehicle is manned by one driver. Service times of v_1 at all pickup and delivery locations are $s_n^{v_1} = 0.75$ hours. The tour of vehicle v_1 will start at time $t^{v_1}_{\text{start}}$. The driver has already been instructed to pick up two ULD of type D in Frankfurt which have to be delivered to Hamburg. Vehicle v_2 has a capacity of 2D2A, i.e. two ULD of type D and two ULD of type A. The cargo body is rigid sided, equipped with a refrigerating unit, and has a roller platform. The vehicle is manned by two drivers. Service times of v_2 at all pickup and delivery locations are $s_n^{v_2} = 1$ hour. The tour of vehicle v_2 started at time $t^{v_2}_{\text{start}}$. The vehicle is already en-route towards Düsseldorf and cannot be diverted before leaving Düsseldorf at time

t^{v_2}. In Düsseldorf the vehicle will pick up two ULD of type D and one ULD of type A which must be delivered to Cologne.

Order o_1 is a request for delivering one ULD of type A from Frankfurt to Cologne. No special requirements are made for order o_1. Order o_2 is a request for delivering one ULD of type C from Frankfurt to Cologne and another from Frankfurt to Hamburg. To speed up loading and unloading a cargo body with tautliners is requested. Order o_3 is a request for delivering three ULD of type A from Cologne to Frankfurt. The shipments of order o_3 are high value electronic components and no additional other shipments may be loaded simultaneously to prevent damage to the load. A double manned vehicle with rigid sided cargo body is requested for security reasons. Furthermore, roller platforms are requested to ease loading and unloading. All orders are confirmed and an external carrier can be employed to serve order $o \in \mathcal{O}$ at a fixed price p_o.

Obviously, not all orders can be served by all vehicles and for each pair $(o,v) \in \mathcal{O} \times \mathcal{V}$ the compatibility parameter δ_{ov} has to be determined. For this purpose, a vector q^v is used representing the properties of vehicle $v \in \mathcal{V}$. The number of drivers is represented by $q^v\big|_{\text{drivers}}$, the parameter $q^v\big|_{\text{canvas sides}} \in \{0,1,2\}$ indicates whether the vehicle has canvas sides or not, and whether tautliners are provided. The binary parameters $q^v\big|_{\text{rigid box}}$, $q^v\big|_{\text{roller platform}}$, and $q^v\big|_{\text{temperature}}$ indicate whether rigid boxes, roller platforms, and refrigerating units are provided. The properties of vehicles $v \in \mathcal{V}$ are represented by

$$q^v = \begin{pmatrix} q^v\big|_{\text{drivers}} \\ q^v\big|_{\text{canvas sides}} \\ q^v\big|_{\text{rigid box}} \\ q^v\big|_{\text{roller platform}} \\ q^v\big|_{\text{temperature}} \end{pmatrix} \Rightarrow q^{v_1} = \begin{pmatrix} 1 \\ 2 \\ 0 \\ 0 \\ 0 \end{pmatrix} \quad q^{v_2} = \begin{pmatrix} 2 \\ 0 \\ 1 \\ 1 \\ 1 \end{pmatrix}.$$

For each order $o \in \mathcal{O}$ the vector q_o represents the minimal requirements of the order. Obviously, an over-fulfillment of the requirements is also possible, e.g. if regular canvas sides are required, a vehicle with tautliners may also serve the transportation request. The requirements of orders $o \in \mathcal{O}$ are represented by

$$q_o = \begin{pmatrix} q_o\big|_{\text{drivers}} \\ q_o\big|_{\text{canvas sides}} \\ q_o\big|_{\text{rigid box}} \\ q_o\big|_{\text{roller platform}} \\ q_o\big|_{\text{temperature}} \end{pmatrix} \Rightarrow q_{o_1} = \begin{pmatrix} 1 \\ 0 \\ 0 \\ 0 \\ 0 \end{pmatrix} \quad q_{o_2} = \begin{pmatrix} 1 \\ 2 \\ 0 \\ 0 \\ 0 \end{pmatrix} \quad q_{o_3} = \begin{pmatrix} 2 \\ 0 \\ 1 \\ 1 \\ 0 \end{pmatrix}.$$

Given the vectors q^v for all $v \in \mathcal{V}$ and q_o for all $o \in \mathcal{O}$ the compatibility parameters δ_{ov} can be determined by

$$\delta_{ov} = 1 \Leftrightarrow q_o \leq q^v.$$

As said before, ULD of type A, C, and D have the same ground area measurements, whereas contour type D is higher than contour type C, which is higher than contour type A. Obviously, a vehicle that can load a certain number of ULD of a type can also load the same number of ULD of a type with smaller height. The capacity of vehicles $v \in \mathcal{V}$ can be represented by

$$r^v = \begin{pmatrix} r^v\big|_{\text{ULD A}} \\ r^v\big|_{\text{ULD C}} \\ r^v\big|_{\text{ULD D}} \\ r^v\big|_{\text{FTL}} \end{pmatrix} \Rightarrow r^{v_1} = \begin{pmatrix} 5 \\ 5 \\ 5 \\ M \end{pmatrix} \quad r^{v_2} = \begin{pmatrix} 4 \\ 2 \\ 2 \\ M \end{pmatrix}$$

whereas, $r^v\big|_{\text{ULD A}}$ represents the maximal number of ULD of type A that can be loaded, $r^v\big|_{\text{ULD C}}$ represents the maximal number of ULD of type C that can be loaded, and $r^v\big|_{\text{ULD D}}$ represents the maximal number of ULD of type D that can be loaded. The parameter $r^v\big|_{\text{FTL}}$ is used to consider restrictions to load consolidation. If no other order must be served simultaneously, an order can be interpreted as a full-truckload (FTL) request, which requires the entire capacity M of a vehicle. All other orders only require a portion of the capacity M. Note that M is any sufficiently large number which does not give any information about the properties of the vehicle.

The resource demand at node $n \in \mathcal{N}$ can be represented by:

$$r_n = \begin{pmatrix} r_n\big|_{\text{ULD A}} \\ r_n\big|_{\text{ULD C}} \\ r_n\big|_{\text{ULD D}} \\ r_n\big|_{\text{FTL}} \end{pmatrix} \Rightarrow$$

$$r_{n_{(v_1,1)}} = \begin{pmatrix} 0 \\ 0 \\ 0 \\ 0 \end{pmatrix} \quad r_{n_{(v_1,2)}} = \begin{pmatrix} 2 \\ 2 \\ 2 \\ 2 \end{pmatrix} \quad r_{n_{(v_1,3)}} = \begin{pmatrix} -2 \\ -2 \\ -2 \\ -2 \end{pmatrix} \quad r_{n_{(v_1,4)}} = \begin{pmatrix} 0 \\ 0 \\ 0 \\ 0 \end{pmatrix}$$

$$r_{n_{(v_2,1)}} = \begin{pmatrix} 3 \\ 2 \\ 2 \\ 3 \end{pmatrix} \quad r_{n_{(v_2,2)}} = \begin{pmatrix} -3 \\ -2 \\ -2 \\ -3 \end{pmatrix} \quad r_{n_{(v_2,3)}} = \begin{pmatrix} 0 \\ 0 \\ 0 \\ 0 \end{pmatrix}$$

$$r_{n_{(o_1,1)}} = \begin{pmatrix} 1 \\ 0 \\ 0 \\ 1 \end{pmatrix} \quad r_{n_{(o_1,2)}} = \begin{pmatrix} -1 \\ 0 \\ 0 \\ -1 \end{pmatrix}$$

$$r_{n_{(o_2,1)}} = \begin{pmatrix} 2 \\ 2 \\ 0 \\ 2 \end{pmatrix} \quad r_{n_{(o_2,2)}} = \begin{pmatrix} -1 \\ -1 \\ 0 \\ -1 \end{pmatrix} \quad r_{n_{(o_2,3)}} = \begin{pmatrix} -1 \\ -1 \\ 0 \\ -1 \end{pmatrix}$$

$$r_{n_{(o_3,1)}} = \begin{pmatrix} 3 \\ 0 \\ 0 \\ M \end{pmatrix} \quad r_{n_{(o_3,2)}} = \begin{pmatrix} -3 \\ 0 \\ 0 \\ -M \end{pmatrix}$$

As can be seen, order o_1 may be served by both vehicles. Order o_2 may only be served by vehicle v_1 due to compatibility constraints. Order o_3 may only be served by vehicle v_2 due to compatibility constraints. Furthermore, o_3 must not be served simultaneously with orders o_1 and o_2 due to capacity constraints. Figure 5.4 shows a solution of the GVRP-DWH which is feasible according to compatibility and capacity constraints.

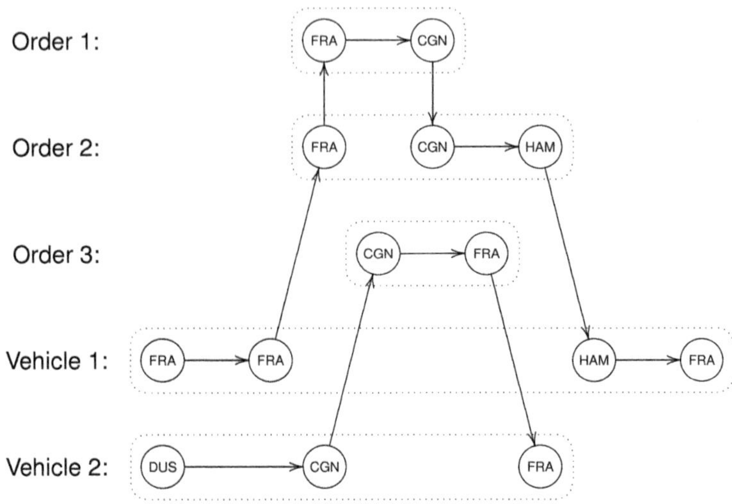

Fig. 5.4: Solution of the GVRP-DWH

For each vehicle $v \in \mathcal{V}$ time windows at $n_{(v,1)}$ are set to $[t^v, t^v]$, whereas t^v denotes the earliest time vehicle v can be diverted from its tour. For each vehicle $v \in \mathcal{V}$ time windows at $n_{(v,\lambda_v)}$ are set to $[t^v, t^v_{\text{end}}]$, whereas t^v_{end} denotes the latest time the tour of vehicle v has to end, i.e. 144 hours after t^v_{start}. The time windows of all other nodes are set according to the restrictions on pickup and delivery times.

For the single-manned vehicle v_1 the parameters required to consider drivers' working hours are set to $t^{v_1}_{\text{weekly}} = 56$ hours, $t^{v_1}_{\text{daily}} = 9$ hours, $t^{v_1}_{\text{nonstop}} = 4.5$ hours, $t^{v_1}_{\text{rest}} = 11$ hours, and $t^{v_1}_{\text{break}} = 0.75$ hours. Vehicle v_2 is manned by two drivers and the parameters required to consider drivers' working hours are set to $t^{v_2}_{\text{weekly}} = 112$ hours, $t^{v_2}_{\text{daily}} = 18$ hours, and $t^{v_2}_{\text{nonstop}} = 18$ hours, and $t^{v_2}_{\text{rest}} = 9$ hours. Note that $t^{v_2}_{\text{break}}$ is not required, as one driver can have his break whilst the other is driving.

6

Dynamic vehicle routing

6.1 Introduction

The construction of schedules is a key issue for motor carriers and computer-based decision support systems have a big impact on the profitability of commercial vehicle operations. If all relevant data is known a priori, schedules can be generated using static planning systems. In many real-life applications, however, relevant data change during the execution of transportation processes and schedules have to be updated dynamically. This chapter investigates the main differences between dynamic and static planning. Algorithms developed for the classical models presented in the previous chapter are surveyed, focusing on those algorithms that are suitable for rich vehicle routing problems in which data may change dynamically. Most algorithms surveyed are based on so-called *neighbourhood operators* that are used to move from one solution in the search space to another. As the GVRP generalises the classical models, not all of these neighbourhood operators can be successfully applied to the GVRP or GVRP-DWH. This chapter introduces neighbourhood operators for the GVRP and the GVRP-DWH. Two insertion methods are presented that can be used to quickly improve a solution considering new transportation requests arriving dynamically. A Reduced Variable Neighbourhood Search algorithm is introduced which achieves its strength from changing the neighbourhood structure during the search. Large Neighbourhood Search algorithms are introduced which iteratively remove some of the transportation requests from the current schedule and then, re-insert them using one of the insertion methods presented. The algorithms presented are characterised by very fast response times and can be used within the Dynamic Planning System presented in section 4.4.3. This chapter concludes by presenting computational experiments performed on test cases generated for the GVRP and GVRP-DWH.

6.2 Dynamic vs. static planning

This section discusses the main differences between dynamic and static vehicle routing. Although some of the issues discussed apply for dynamic and static planning, their impact on dynamic planning is often much more severe. A comprehensive discussion of dynamic vehicle routing can be found in Psaraftis (1988) and Psaraftis (1995). In brief, Psaraftis gives the following definition of a dynamic problem: a problem is *dynamic* if information on the problem is made known to the decision maker or is updated concurrently with the determination of the solution. By contrast, if all inputs are received before the determination of the solution and do not change thereafter, the problem is termed *static*[1].

6.2.1 Evolution of information

Obviously, the major difference between dynamic and static problems is the evolution of information. In static problems information is assumed to be known for the entire duration of the transportation process. In dynamic problems, however, some input is not known at the time of planning, and some input is not known with certainty. For example, traffic conditions change, the number of vehicles available may change due to vehicle break-down, new transportation requests become known, or attributes of transportation requests may change.

In the collection of fresh milk, for example, pickup locations are generally known in advance and do not change dynamically. However, the amount of milk to be collected may change significantly and is only known with certainty when the vehicle reaches the pickup location. Courier companies face a different challenge: not all customer locations are known when the vehicles start their tours and new transportation requests arrive while the vehicles are en-route.

Most of the literature on dynamic vehicle routing only considers one type of uncertainty: the arrival of a new transportation request. Larsson (2000) and Kilby et al. (1998) measure the degree of dynamism by the number of new transportation requests arriving during the planning process. Few work consider problems with several sources of uncertainty. Among them are the works by Powell et al. (2000) and Fleischmann et al. (2004), who consider the arrival of new transportation requests and uncertain travel times.

Another issue in dynamic planning is the reliability of future information. Long-term information is more likely to change than short-term information. For example, long-term estimations of travel times can only be based on historical data. In the short-term, however, actual traffic conditions can be used to estimate travel times. Those short-term forecasts allow a much better estimation of what is going to happen and thus, help in improving punctuality.

[1] In this definition the term *problem* refers to the problem defined by the corresponding analytical model and not the real-life problem which generally is dynamic anyway.

6.2.2 Rolling horizon

In static planning, schedules are generated for a certain finite planning horizon. The duration of the planning process is bounded by the time between data collection and the start of transportation processes. Given the input data, a static planning system must be capable of calculating high quality solutions in the time available for optimisation.

In dynamic planning, the planning horizon may neither be bounded, nor known. In fact, a typical dynamic planning scenario is that of an open-ended process, going on for an indefinite period of time. Usually, near-term events are more important than long-term events. The consideration of requirements, which have to be satisfied way into future, would not be very wise, because such future information may change anyway. In a typical rolling horizon framework only information relevant to planning decisions within a horizon of a certain length is considered. As time unfolds, parts of the tentative schedule are applied and new information may enter the planning horizon.

6.2.3 Impreciseness of model representation

Real-life vehicle routing problems usually cannot be precisely represented by an analytical model which is required for computer-based decision support. Even if the analytical model is of high quality, discrepancies between model representation and real-life problem arise as a result of the sheer cost of getting information into the computer[1]. Telematics systems can be used to improve the timely availability of information regarding the actual transportation processes. Electronic Data Interchange (EDI) can be used to integrate information systems of shippers, e.g. to obtain all relevant data regarding transportation requests and customer locations. Despite of the improved possibilities of getting the data into the model, the information is generally not only incomplete but also imprecise. A shipper, for example, may ask that a shipment be picked up in the morning before noon, when his dock is not as busy. In the model such restrictions are usually treated as time window constraints. The computer system has no way of interpreting whether such a request for early pickup is a hard constraint or whether the shipper was only trying to express a preference.

As illustrated in figure 6.1, the impreciseness of the model representation results in two fundamental problems:

- some solutions which are feasible according to the model may not be feasible in reality and vice versa,
- a solution with high quality in the model may not have the same high quality in reality.

Although these problems occur in static as well as in dynamic planning, the impact is quite different. In static planning there is usually more time for the collection of data, resulting in a more accurate representation of the real-life problem. Furthermore,

[1] See Powell et al. (2002)

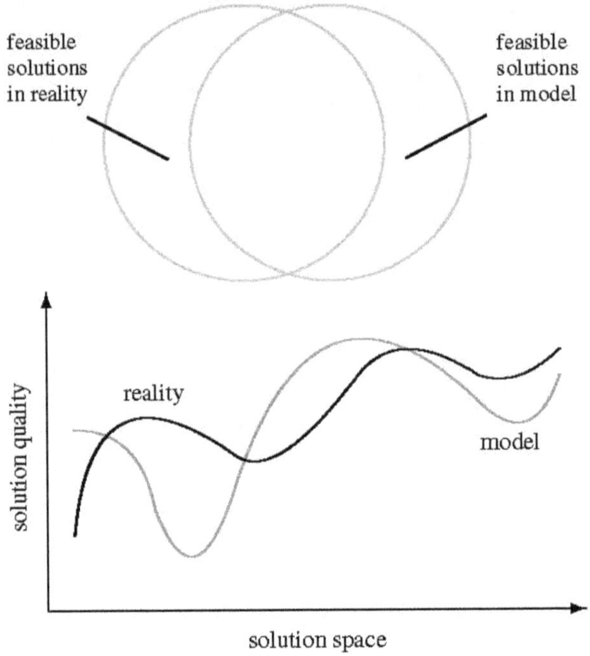

Fig. 6.1: Differences between real-life problem and model

there is more time to manually verify and validate a solution recommended by the planning system.

6.2.4 Interactivity

Due to the impreciseness of any model representation and the fact that a significant amount of relevant information is not available to the computer, but only to the dispatchers who are in direct contact with drivers and shippers, model recommendations cannot always be fully implemented. According to Powell et al. (2000), several motor carriers report that the average usage of model recommendations is below 60%, and good performance is considered around 70%.

In order to deal with this issue, Kopfer and Schönberger (2002) have presented a framework for interactive problem solving. The concept is founded in posting a problem to a planning method in order to let it generate a solution. The returned solution usually does not satisfy all real-life requirements. Therefore, dispatchers may add, modify, or remove certain constraints in the analytical model. The modified problem is again posted to the planning method, and after a solution has been found, further modifications to the model can be made. Although this approach can be used for static and dynamic problems, this iterative decision making is much harder in dynamic planning due to the lack of time.

A similar approach has been presented in section 4.4.3. Instead, of an iterative decision making process, the real-time decision support system presented in section 4.4.3 allows dispatchers and dynamic planning system to simultaneously modify the current solution. Dispatchers may add, modify, or remove certain constraints in the analytical model at any time. All changes made by the dispatchers are directly considered by the dynamic planning system which continuously searches for improving solutions. Whenever the dynamic planning system finds a better solution, the current solution is immediately updated and shown to the dispatchers. An optimistic locking scheme is used in order to maintain data consistency.

Humans use a form of cache memory for processing information, called *working memory*[1]. This working memory must be regularly updated in order to consider changes in problem data and solution. A dynamic planning system must support dispatchers in quickly updating their working memory. Therefore, besides of providing algorithmic optimisation techniques, a dynamic planning system must also provide sophisticated graphical user interfaces (GUI) allowing dispatchers to quickly identify modified parts of the solution, and to efficiently verify feasibility and profitability of an automatically generated schedule.

6.2.5 Response time

The *response time* of an algorithm is the time which is needed until a newly calculated solution can be applied. Algorithms for dynamic planning must have fast response times for two reasons. First, a solution calculated for a dynamic problem can only be applied if the input data have not changed significantly during the planning process. Second, the longer it takes to calculate a new solution the higher is the probability that dispatchers concurrently change the current solution.

In many cases dispatchers have to decide about load acceptance or rejection while the shipper is on the phone making the request. Dynamic planning systems should be capable of supporting dispatchers in the load acceptance decision while the shipper is on the phone. That is, within a couple of seconds they should give support in deciding whether a transportation request can be served feasibly and efficiently.

The main advantage in dynamic planning is that there is usually plenty of time for optimisation. As noted by Kilby et al. (1998), ten minutes spent to find a solution for a small problem may seem like a long time in the static case. In the dynamic case, however, one has all day so the time might as well be used. A dynamic planning system can be used to successively improve the current solution. Obviously, fast response times can not be achieved if new solutions are calculated from scratch every time the dynamic planning method is invoked. Therefore, a dynamic planning system should have a *restart* capability, i.e. the planning system should be able to continue from the current solution. Furthermore, dynamic planning methods require efficient information update mechanisms in order to efficiently consider modified input data.

[1] See Powell et al. (2002)

6.2.6 Measuring performance

Measuring performance in dynamic planning is much more complicated than in static planning. A method for static planning can be evaluated by comparing the solution obtained with solutions obtained by other methods. In dynamic planning, however, the decisions made at one point in time determine the options and alternatives at a later point in time. It is possible to compare alternative decisions at a point in time, but then carrying the effects of those decisions forward in time creates a problem[1].

In dynamic planning with rolling horizons any evaluation of operating efficiencies must address the issue of time period evaluation[2]. The state of the system at the end of the evaluation period may have a dramatic impact on future performance. A dynamic planning method may perform quite well over a day or a week, but can have poor long-term performance, e.g. if all vehicles are sent to very distant regions where there is little hope of picking up a new load.

As mentioned above, dynamic planning methods must be interactive enabling dispatchers to verify model recommendations. In many cases dispatchers will agree with model recommendations. However, different problem knowledge and solution methods of computer and human dispatchers may result in contradicting decisions. Under the time constraints in dynamic planning it is often very difficult or even impossible to find out why the recommendation is being made. Is the discrepancy a result of "higher reasoning" or a simple data error? Typically, dispatchers will follow their own intuition and not the model recommendation. As noted by Powell et al. (2002), a dynamic planning system is often considered successful if dispatchers agree with model recommendations, but dispatchers often do not follow model recommendations. Obviously, the question arises that if dispatchers do not comply with model recommendations are the solutions produced good at all?

6.3 State of the art

Since the first formulation of the VRP by Dantzig and Ramser (1959), thousands of algorithms have been proposed for the optimal and approximate solution of the VRP, PDP and their variants. The vast majority of these algorithms concerns the static version of the problems, i.e. it is assumed that all input data is known and invariant during optimisation. Only recently dynamic problems have been increasingly studied in the vehicle routing literature. This section gives an overview of algorithms for solving vehicle routing problems. The survey focuses on solution methods which can be used for rich problems in which problem data may change dynamically. A full survey of methods for the VRP, the PDP, and their variants would be out of scope of this work and the reader is referred to the survey on the PDP by Mitrović-Minić (1998), the book on the VRP and its variants edited by Toth and Vigo (2002), as well as the more recent survey on the VRP by Cordeau et al. (2004) and secondary literature

[1] See Powell et al. (2002)

[2] See Regan et al. (1998)

given there. The solution methods discussed in this section can be categorised into *assignment methods*, *construction methods*, *improvement methods*, *meta-heuristics* and *mathematical programming based methods.*

6.3.1 Assignment methods

Assignment methods are methods that assign transportation requests to vehicles for immediate execution. They are used in highly dynamic problems where problem data change very fast and no foresighted planning is likely to perform well. Assignment methods can be used for the VRP and the Full-Truckload PDP, i.e. the special case of the PDP in which all transportation requests are FTL requests[1]. Assignment rules for the Full-Truckload PDP that either assign a newly arrived order to an idle vehicle or a vehicle which just becomes available to an open order have been presented by Regan et al. (1998) and more recently by Fleischmann et al. (2004). Assignment algorithms that simultaneously assign several open orders to idle vehicles are studied by Spivey and Powell (2004) and Fleischmann et al. (2004).

6.3.2 Construction methods

Construction methods gradually build tours while keeping an eye on the objective function value, but they do not contain an improvement phase per se, see Laporte and Semet (2002). A comprehensive survey on construction methods for the VRPTW is given by Bräysy and Gendreau (2005a). One of the best-known tour construction methods for the VRP is the savings algorithm by Clarke and Wright (1964). The algorithm starts by creating a tour for each order and successively merges two tours to form a new tour replacing the former. The merge is made according to the maximum savings in the objective function. This algorithm naturally applies to problems for which the number of vehicles is not fixed.

Insertion methods are methods that successively insert open transportation requests into partially constructed tours. They are well suited for dynamic planning, because they permit to incorporate a new order considering the set of tours which are currently implemented. Insertion methods are very fast and can be used for dynamic vehicle routing problems in which there may not be enough time to employ more sophisticated methods. Furthermore, insertion methods can often be applied to problems incorporating various real-life requirements without loosing efficiency. A discussion of efficient insertion methods for vehicle routing problems incorporating complicating constraints can be found in Campbell and Savelsbergh (2004).

Early examples of insertion methods have been proposed by Wilson et al. (1970) for the DARP and by Solomon (1987) for the VRPTW. Parallel insertion methods for the static VRPTW which simultaneously construct several tours via insertions are proposed by Potvin and Rousseau (1993) and Antes and Derigs (1995). Shen et al. (1995) presented a computer assistant whose aim is to help dispatchers in inserting new transportation requests into existing tours for the PDP. The computer assistant

[1] See section 3.3.3.1

integrates a learning module based on neural networks which is trained by an expert dispatcher. A graphical user interface allows the dispatchers to make the final insertion decision. Recently Lu and Dessouky (2006) presented an insertion method for the PDPTW which not only considers the classical incremental costs but also the cost of reducing the time window slack so that more opportunities are left for future insertions. Insertion methods for the dynamic PDP are also studied by Yang et al. (1999), Mitrović-Minić (2001) and Fleischmann et al. (2004).

6.3.3 Improvement methods

Many solution techniques for combinatorial optimisation problems are based on a simple and general idea. Let s be a feasible solution of the problem considered and let $f(s)$ denote the objective function value of s. For each feasible solution s the *neighbourhood* of s is defined by the solutions s^* which can be obtained by applying an appropriately defined neighbourhood operator to the solution s. So-called *local search* or *neighbourhood search* methods explore the neighbourhood of the current solution s by searching for a feasible solution s^* of high quality in the neighbourhood of the current solution s. This solution may be accepted as new current solution, and in this case, the process is iterated by considering s^* as new current solution[1].

In maximisation (minimisation) problems, a new solution s^* is typically only accepted if $f(s^*) \geq f(s)$ $\big(f(s^*) \leq f(s)\big)$. If no solution s^* with $f(s^*) > f(s)$ $\big(f(s^*) < f(s)\big)$ exists in the neighbourhood of s, a *local optimum* has been reached. A comprehensive work on local search methods is given by Aarts and Lenstra (1997). A survey and comparison of local search methods for the VRPTW has been presented by Bräysy and Gendreau (2005a).

Improvement methods are local search methods which start with a feasible solution and gradually modify the current solution in order to improve the solution quality. The most simple improvement methods operate on a single tour and optimise the sequence in which the locations are visited. They are often based on methods developed for the TSP, e.g. 2-opt by Lin (1965) and Or-opt by Or (1976). Others consider several tours simultaneously e.g. the operators relocate, exchange, and cross originally proposed by Savelsbergh (1992) for the classical VRP. Local optima produced by an improvement methods can be very far from the optimal solution, as they only accept solutions that produce an improvement in the objective function value. Thus, the outcome heavily depends on the initial solution and the neighbourhood definition.

6.3.4 Meta-heuristics

Meta-heuristics are general solution procedures that often embed some of the standard tour construction and improvement methods, as well as techniques to escape from local optima of low quality, see Gendreau et al. (2002). A comprehensive survey on the use of meta-heuristics for the VRPTW is given by Bräysy and Gendreau

[1] Note that insertion methods can also be interpreted as local search or neighbourhood search methods if not all transportation requests must be served.

(2005b). Examples of meta-heuristics are Simulated Annealing, Genetic Algorithms, Ant Systems, Tabu Search, and Iterated Local Search, see e.g. Blum and Roli (2003).

The fundamental idea of Simulating Annealing is to allow moves resulting in solutions of worse quality in order to escape from locally optimal solutions, see Kirkpatrick et al. (1983). The probability of doing such a move is decreased during the search. Although successful for many static problems, it is not clear how to effectively change this probability in dynamic problems, as input data may change during the search.

Genetic Algorithms, Ant Systems, and Tabu Search are memory-based methods classified as *Adaptive Memory Programming* (AMP) methods by Taillard et al. (1998). Particularly in highly dynamic problems, AMP methods require methods to efficiently update the memory. The memory can only be used effectively if there are only minor changes to the problem data. Examples for AMP methods are the Genetic Algorithm for the dynamic PDP presented by Pankratz (2004), an Ant Colony System for the dynamic VRP by Montemanni et al. (2002), the Tabu Search algorithm for the dynamic VRP by Gendreau et al. (1999), and the Tabu Search algorithms for the dynamic PDP by Gendreau et al. (1998) and Mitrović-Minić (2001).

The essence of Iterated Local Search (ILS) is to iteratively build a sequence of solutions generated by an embedded heuristic. It applies the heuristic until it finds a local optimum. Then it perturbs the solution and restarts the heuristic. This generally leads to far better solutions than if one would use repeated random trials of that heuristic, see Lourenço et al. (2002).

Variable Neighbourhood Search (VNS) can be interpreted as a specialised ILS based on the idea of systematically changing the neighbourhood structure during the search, see Mladenović and Hansen (1997) and Hansen and Mladenović (2003). VNS systematically exploits the following observations: a) a local optimum with respect to one neighbourhood structure is not necessary so for another; b) a global optimum is a local optimum with respect to all possible neighbourhood structures; c) for many problems local optima with respect to one or several neighbourhoods are relatively close to each other. An example of a VNS algorithm for vehicle routing problems is the algorithm for the multi-depot VRPTW presented by Polacek et al. (2004).

As noted by Ahuja et al. (2002), a critical issue in the design of a neighbourhood search approach is the size of the chosen neighbourhood. Large neighbourhoods increase the quality of the locally optimal solutions, however, locally optimal solutions are difficult to find in very large neighbourhoods. In each iteration of the Large Neighbourhood Search (LNS) algorithm presented by Shaw (1997) for the VRPTW, k customers are first removed from their tours and then, re-inserted using a branch and bound procedure. Schrimpf et al. (2000) and Ropke and Pisinger (2006) present similar LNS algorithms using fast insertion heuristics for the re-insertion of transportation requests. The use of fast insertion heuristics is more appropriate for dynamic planning as fast response times can be easily achieved. The LNS approach is very well suited for highly constrained vehicle routing problems, see Kilby et al. (2000), and rich vehicle routing problems in which data may change dynamically, see Goel and Gruhn (2007).

6.3.5 Mathematical programming based methods

Finding optimal solutions for vehicle routing problems is generally too time consuming, particularly in dynamic planning. However, *mathematical programming based methods* can be used in dynamic vehicle routing if the number of decision variables is reduced dramatically or subproblems are only solved approximately.

Mathematical programming based methods for the dynamic Full-Truckload PDP have been presented by Yang et al. (1999; 2004). To guarantee robustness and timeliness of the methods, the number of transportation requests involved in each optimisation was limited to a fixed upper-bound. Thus, the resulting mathematical program is significantly reduced in size and is solved using a branch-and-cut procedure in the CPLEX solver[1].

The most popular mathematical programming based methods for rich vehicle routing problems is the Column Generation approach. Column Generation has been applied to the VRPTW by Desrochers et al. (1992), the HFVRP by Taillard (1996), the PDP by Dumas et al. (1991), a generalised PDP by Savelsbergh and Sol (1998), and a rich PDP considering multiple time windows and drivers' working hours as regulated by the U.S. Department of Transportation by Xu et al. (2003). Column Generation approaches for dynamic vehicle routing problems have been presented by Savelsbergh and Sol (1998) and more recently by Chen and Xu (2006).

6.4 Neighbourhood operators

Neighbourhood operators are used to move from one solution s in the search space to another solution s^* in the neighbourhood of s. They are the core of most heuristics and are typically defined in a way that they are easy to calculate and to evaluate. As the GVRP generalises the classical models, not all neighbourhood operators used by the solution methods surveyed in the previous section can be successfully applied to the GVRP. In particular the arc based operators 2-opt[2], OR-opt[3], or cross[4] are unlikely to perform well for the GVRP as they are likely to produce solutions in which precedence or grouping constraints imposed on the GVRP are violated. Operators used for the PDP such as PD-rearrange, PD-shift, and PD-exchange, presented by Li and Lim (2001), can be easily modified in order to work for the GVRP. As not all transportation requests must be served in the GVRP, neighbourhood operators that change the number of transportation requests covered by the solution can also be applied to the GVRP.

In order to move from one feasible solution to another using neighbourhood operators, an initial feasible solution must be available. Under the assumption

$$\text{for all vehicles } v \in \mathcal{V}: \quad \theta_v := (n_{(v,1)}, \ldots, n_{(v,\lambda_v)}) \text{ is a feasible tour} \tag{A}$$

[1] See e.g. ILOG (2001)

[2] See Lin (1965)

[3] See Or (1976)

[4] See Savelsbergh (1992)

a feasible solution of the GVRP can naturally be obtained by setting the tour of each vehicle $v \in \mathcal{V}$ to $(n_{(v,1)}, \ldots, n_{(v,\lambda_v)})$. Therefore, finding an initial feasible solution is trivial in the GVRP and throughout this work it is assumed that assumption (A) is satisfied.

6.4.1 INSERT

The INSERT operator is the basis of all insertion methods and inserts all locations belonging to an unscheduled order into the tour of a vehicle, subject to compatibility and precedence constraints imposed on the GVRP.

Definition 13 *An insertion of an order $o \in \mathcal{O} \setminus \mathcal{O}_\theta$ to a tour θ is a transformation of θ to a tour θ^* such that θ is a subsequence of θ^* with $\mathcal{O}_{\theta^*} = \mathcal{O}_\theta \cup \{o\}$. An insertion is feasible if and only if the resulting tour is feasible.*

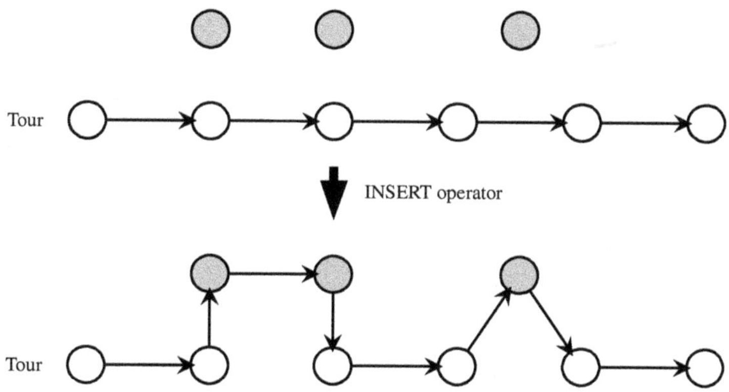

Fig. 6.2: INSERT

Figure 6.2 illustrates the insertion of an unscheduled order to the tour of a vehicle. INSERT-moves can be easily evaluated considering the revenue of the inserted order and the incremental costs for vehicle movements. Let θ and θ^* denote the initial tour and the tour after insertion of the previously unscheduled order $o \in \mathcal{O}$. Let $\mathcal{A}_\theta$ and $\mathcal{A}_{\theta^*}$ denote the arcs used by θ and θ^*. The *incremental cost* of an insertion is the difference between the costs of the new tour θ^* and the tour θ.

$$\begin{aligned}
\text{incremental costs} &= c_{\theta^*} - c_\theta \\
&= \sum_{(n,m)\in\mathcal{A}_{\theta^*}} c^v_{nm} - \sum_{(n,m)\in\mathcal{A}_\theta} c^v_{nm} \\
&= \sum_{(n,m)\in\mathcal{A}_{\theta^*}\setminus\mathcal{A}_\theta} c^v_{nm} - \sum_{(n,m)\in\mathcal{A}_\theta\setminus\mathcal{A}_{\theta^*}} c^v_{nm}
\end{aligned}$$

As we can see, it suffices to determine the difference between costs for vehicle movements on new arcs reduced by the costs for vehicle movements on arcs which are no longer used.

The *efficiency* of an insertion is the difference between the revenue of the inserted order and the incremental cost of the insertion.

$$\text{efficiency} = p_o - (c_{\theta^*} - c_\theta)$$

A positive efficiency implies that the new solution obtained by applying the INSERT-move is better than the previous solution.

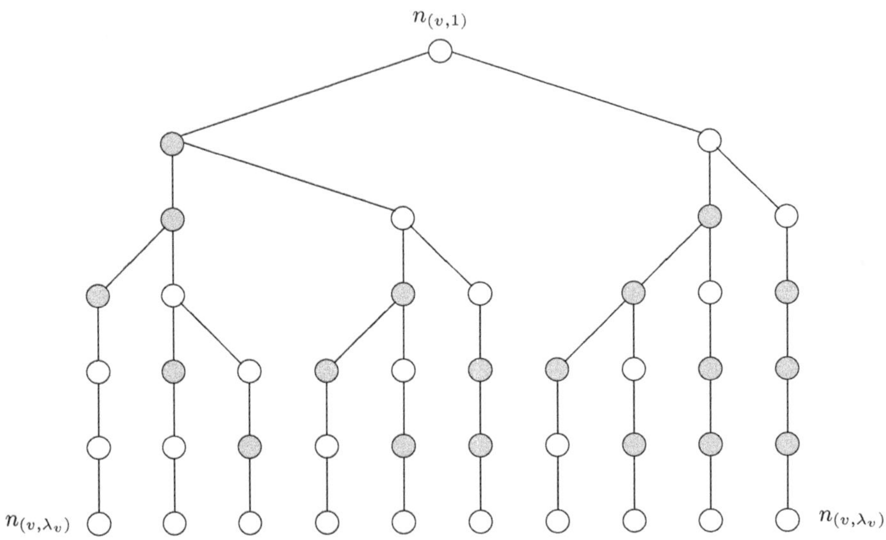

Fig. 6.3: Insertion tree

To explore the neighbourhood defined by the INSERT operator an *insertion tree* can be constructed for each tour θ and each unscheduled order $o \in \mathcal{O}$. The root of the insertion tree corresponds to the starting point of the tour. At each node the insertion tree has at most two branches: one for a possible insertion of the next location belonging to order o and one for the next node of the tour θ. Each branch is continued (with all necessary subbranches) to complete a tour covering the order. Figure 6.3 shows an example of an insertion tree of an order $o \in \mathcal{O}$ with $\lambda_o = 3$ and a tour θ with $\text{length}(\theta) = 4$. The white nodes represent the nodes of tour θ, the gray nodes represent the nodes corresponding to order o.

Obviously, not all of the $\binom{\lambda_o + \text{length}(\theta) - 2}{\lambda_o}$ possible INSERT-moves will result in a feasible tour. If infeasibility can be detected at a branch in the insertion tree, the branch can be cut off and does not have to be explored. To achieve good performance using the INSERT operator, it is desirable that feasibility of insertions can be determined quickly.

Suppose we have generated tour $\theta^* := (m_1, \dots, m_k)$ by inserting order $o \in \mathcal{O}$ to a feasible tour $\theta := (n_1, \dots, n_h)$. Then, there exists an index $i \geq 0$ with $(n_{h-i}, \dots, n_h) = (m_{k-i}, \dots, m_k)$. Now, assume that for all nodes in the sequence $(m_1, \dots, m_{k-i-1})$ capacity and time window constraints are satisfied.

If $\sum_{\mu=1}^{\mu \leq \lambda_o} r_{n_{(o,\mu)}} = 0$ we know that capacity constraints are also satisfied for all nodes in the sequence $(m_{k-i}, \dots, m_k)$. If $\sum_{\mu=1}^{\mu \leq \lambda_o} r_{n_{(o,\mu)}} \neq 0$ capacity constraints may be violated for nodes in the sequence $(m_{k-i}, \dots, m_k)$. Let $\rho_{n_j} := \sum_{l=1}^{l \leq j} r_{n_l}$, $\rho_{n_h}^{\max} := \rho_{n_h}$, $\rho_{n_h}^{\min} := \rho_{n_h}$ and for $j < h$ let

$$\rho_{n_j}^{\max} := \text{element wise maximum of } \rho_{n_j} \text{ and } \rho_{n_{j+1}}^{\max}$$

and

$$\rho_{n_j}^{\min} := \text{element wise minimum of } \rho_{n_j} \text{ and } \rho_{n_{j+1}}^{\min}.$$

Lemma 1 *The tour $\theta^* := (m_1, \dots, m_k)$ is feasible according to capacity constraints if and only if*

$$\sum_{\mu=1}^{\mu \leq \lambda_o} r_{n_{(o,\mu)}} \leq r^v - \rho_{n_{h-i}}^{\max}$$

and

$$\sum_{\mu=1}^{\mu \leq \lambda_o} r_{n_{(o,\mu)}} \geq -\rho_{n_{h-i}}^{\min}.$$

Proof: For $i = 0$ the lemma is trivial. Now, suppose these inequalities are satisfied. Then, we have for $j \in \{0, \dots, i\}$

$$\sum_{l=1}^{l \leq k-j} r_{m_l} = \rho_{n_{h-j}} + \sum_{\mu=1}^{\mu \leq \lambda_o} r_{n_{(o,\mu)}} \leq \rho_{n_{h-j}}^{\max} + r^v - \rho_{n_{h-i}}^{\max} \leq r^v$$

and

$$\sum_{l=1}^{l \leq k-j} r_{m_l} = \rho_{n_{h-j}} + \sum_{\mu=1}^{\mu \leq \lambda_o} r_{n_{(o,\mu)}} \geq \rho_{n_{h-j}}^{\min} - \rho_{n_{h-i}}^{\min} \geq 0.$$

As capacity constraints are satisfied for all nodes in the sequence $(m_1, \dots, m_{k-i-1})$ we have

$$0 \leq \sum_{l=1}^{l \leq j} r_{m_j} \leq r^v \text{ for all } 1 \leq j \leq k.$$

Now, assume that

$$\sum_{\mu=1}^{\mu \leq \lambda_o} r_{n_{(o,\mu)}} \nleq r^v - \rho_{n_{h-i}}^{\max}.$$

Then, we have for $j \in \{0, \dots, i\}$

$$\sum_{l=1}^{l \leq k-j} r_{m_l} = \rho_{n_{h-j}} + \sum_{\mu=1}^{\mu \leq \lambda_o} r_{n_{(o,\mu)}} \not\leq \rho_{n_{h-j}} + r^v - \rho^{\max}_{n_{h-i}}.$$

That is, there exists at least one element for which the left hand side value is greater than the right hand side value. Due to the definition of $\rho^{\max}_{n_{h-i}}$, there exists an index $j^* \in \{0, \dots, i\}$ such that $\rho_{n_{h-j^*}}$ and $\rho^{\max}_{n_{h-i}}$ have the same value for this element. Thus, the right hand side value equals the capacity of the vehicle for this element, and is exceeded by the left hand side value. Analogously, it can be proven that capacity constraints are violated if

$$\sum_{\mu=1}^{\mu \leq \lambda_o} r_{n_{(o,\mu)}} \not\geq -\rho^{\min}_{n_{h-i}}.$$

Q.E.D.

For the VRPTW Solomon (1987) has given a feasibility criterion concerning time window constraints which can also be applied to the GVRP. Let $t_{n_1}, \dots, t_{n_h}$ denote the arrival times of tour θ. The arrival times of θ^* are calculated by $t^*_{m_1} := t^{\min}_{m_1}$ and for all $1 < j \leq k$ by

$$t^*_{m_j} = \max(t^*_{m_{j-1}} + d^v_{m_{j-1}m_j}, t^{\min}_{m_j}).$$

For $j < h$ let the *slack* time be defined by

$$\sigma_{n_j} := t_{n_{j+1}} - (t_{n_j} + d^v_{n_j n_{j+1}}).$$

Let the *push forward* be defined by $\pi_{n_h} := t^{\max}_{n_h} - t_{n_h}$ and for $j < h$ by

$$\pi_{n_j} := \min(t^{\max}_{n_j} - t_{n_j}, \sigma_{n_j} + \pi_{n_{j+1}}).$$

Lemma 2 *The tour $\theta^* := (m_1, \dots, m_k)$ is feasible according to time window constraints if and only if*

$$t^*_{m_{k-i}} \leq t_{n_{h-i}} + \pi_{n_{h-i}}.$$

Proof: For $i = 0$ the lemma is trivial. Now, suppose we have

$$t^*_{m_{k-j}} \leq t_{n_{h-j}} + \pi_{n_{h-j}}$$

with $j \in \{0, \dots, i\}$. Then, it is easy to see that $t^*_{m_{k-j}} \leq t^{\max}_{m_{k-j}}$. Furthermore, we either have $t^*_{m_{k-j+1}} = t^{\min}_{m_{k-j}}$ and feasibility according to time windows is trivial to see, or we have

$$\begin{aligned}
t^*_{m_{k-j+1}} &= t^*_{m_{k-j}} + d^v_{m_{k-j}m_{k-j+1}} \\
&\leq t_{n_{h-j}} + \pi_{n_{h-j}} + d^v_{m_{k-j}m_{k-j+1}} \\
&\leq t_{n_{h-j}} + \sigma_{n_{h-j}} + \pi_{n_{h-j+1}} + d^v_{m_{k-j}m_{k-j+1}} \\
&= t_{n_{h-j}} + t_{n_{h-j+1}} - (t_{n_{h-j}} + d^v_{n_{h-j}n_{h-j+1}}) + \pi_{n_{h-j+1}} + d^v_{m_{k-j}m_{k-j+1}} \\
&= t_{n_{h-j+1}} + \pi_{n_{h-j+1}}
\end{aligned}$$

Therefore, the same condition is satisfied for $j \leftarrow j+1$, and it is easy to see that tour θ^* is feasible according to time window constraints.

Assume we have $t^*_{m_{k-j}} > t_{n_{h-j}} + \pi_{n_{h-j}}$ with $0 < j \leq i$. Obviously, the tour is infeasible according to time windows if $t^*_{m_{k-j}} > t^{\max}_{m_{k-j}}$. Otherwise, we have $t^{\max}_{n_{h-j}} = t^{\max}_{m_{k-j}} \geq t^*_{m_{k-j}} > t_{n_{h-j}} + \pi_{n_{h-j}}$ and $\pi_{n_{h-j}} = \sigma_{n_{h-j}} + \pi_{n_{h-j+1}}$. Thus, we have

$$\begin{aligned} t^*_{m_{k-j+1}} &\geq t^*_{m_{k-j}} + d^v_{m_{k-j}m_{k-j+1}} \\ &> t_{n_{h-j}} + \pi_{n_{h-j}} + d^v_{m_{k-j}m_{k-j+1}} \\ &= t_{n_{h-j}} + \sigma_{n_{h-j}} + \pi_{n_{h-j+1}} + d^v_{n_{h-j}n_{h-j+1}} \\ &= t_{n_{h-j+1}} + \pi_{n_{h-j+1}} \end{aligned}$$

Therefore, the same condition is satisfied for $j \leftarrow j+1$, and it is easy to see that there exists a $j \in \{0, \ldots, i\}$ with $t^*_{m_{k-j}} > t^{\max}_{m_{k-j}}$. **Q.E.D.**

With these criteria the feasibility of an insertion can be determined in constant time after the last node belonging to the order is inserted to the tour. Unfortunately, lemma 2 cannot be used for the GVRP-DWH as travel times change when breaks and rest periods are scheduled differently. Therefore, feasibility of insertions in the GVRP-DWH can only be determined if the arrival times of the entire tour are updated.

6.4.2 REMOVE

The REMOVE operator is the inverse of the INSERT operator and removes all locations belonging to a scheduled order from a tour.

Definition 14 *A removal of an order $o \in \mathcal{O}_\theta$ from a tour θ is a transformation of θ to a tour θ^* such that θ^* is a subsequence of θ with $\mathcal{O}_{\theta^*} = \mathcal{O}_\theta \setminus \{o\}$. A removal is feasible if and only if the resulting tour is feasible.*

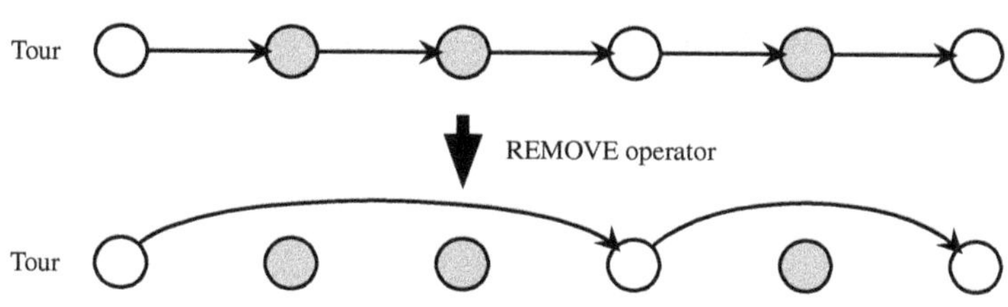

Fig. 6.4: REMOVE

Figure 6.4 illustrates the removal of a transportation request from the tour of a vehicle. It can be evaluated analogously to the INSERT operator. The *savings* of a removal is the difference between the costs of tour θ and the new tour θ^*.

$$\begin{aligned}\text{savings} &= c_\theta - c_{\theta^*} \\ &= \sum_{(n,m)\in\mathcal{A}_\theta} c^v_{nm} - \sum_{(n,m)\in\mathcal{A}_{\theta^*}} c^v_{nm} \\ &= \sum_{(n,m)\in\mathcal{A}_\theta\setminus\mathcal{A}_{\theta^*}} c^v_{nm} - \sum_{(n,m)\in\mathcal{A}_{\theta^*}\setminus\mathcal{A}_\theta} c^v_{nm}\end{aligned}$$

The *efficiency* of a removal is the difference between the savings of the removal and the revenue of the removed order.

$$\text{efficiency} = (c_\theta - c_{\theta^*}) - p_o$$

A positive efficiency implies that the new solution obtained by applying the REMOVE-move is better than the previous solution.

In contrast to the VRP and the PDP not all removals in the GVRP are feasible, as transportation requests can be composed of any mix of pickups and deliveries.

Lemma 3 *Under assumption (A) every removal of an order $o \in \mathcal{O}_\theta$ from a feasible tour θ is feasible if*

$$\sum_{\mu=1}^{\mu\le i} r_{n_{(o,\mu)}} \ge 0 \textit{ for all } o \in \mathcal{O}, 1 \le i \le \lambda_o \tag{C1}$$

or

$$\sum_{\mu=1}^{\mu\le i} r_{n_{(o,\mu)}} \le 0 \textit{ for all } o \in \mathcal{O}, 1 \le i \le \lambda_o. \tag{C2}$$

Proof. It is easy to see that, due to the triangle inequality for travel times, every subsequence of a feasible tour is feasible according to time windows. It remains to proof that capacity constraints are not violated. Let $\theta^* = (n_1, n_2, \ldots, n_k)$ be a tour obtained by removing an order $o \in \mathcal{O}_\theta$ from the tour θ and let i be any point in this tour with $1 \le i \le k$. The accumulated load at n_i is $\rho_{n_i} = \sum_{j=1}^{j\le i} r_{n_j}$. Let

$$i_v = \max\left\{\mu \mid 1 \le \mu \le \lambda_v,\ n_{(v,\mu)} \in \{n_1, \ldots, n_i\}\right\}.$$

For all orders $o \in \mathcal{O}$ with $\{n_{(o,1)}, \ldots, n_{(o,\lambda_o)}\} \cap \{n_1, \ldots, n_i\} \ne \emptyset$ let

$$i_o = \max\left\{\mu \mid 1 \le \mu \le \lambda_o,\ n_{(o,\mu)} \in \{n_1, \ldots, n_i\}\right\}$$

and for all other orders $o \in \mathcal{O}$ let $i_o = 0$ otherwise.

According to assumption (A) we have

$$0 \le \sum_{\mu=1}^{\mu\le j} r_{n_{(v,\mu)}} \le r^v$$

for all $1 \le j \le \lambda_v$. If (C1) holds we have

$$\rho_{n_i} = \sum_{j=1}^{j\leq i} r_{n_j} = \underbrace{\sum_{\mu=1}^{\mu\leq i_v} r_{n_{(v,\mu)}}}_{\geq 0} + \sum_{o\in\mathcal{O}} \underbrace{\sum_{\mu=1}^{\mu\leq i_o} r_{n_{(o,\mu)}}}_{\geq 0} \geq 0.$$

Furthermore, we have $\rho_{n_i} \leq r^v$ as the accumulated load at every point in the tour θ^* is due to (C1) less than or equal to the accumulated load at the corresponding point in the feasible tour θ. If (C2) holds we have

$$\rho_{n_i} = \sum_{j=1}^{j\leq i} r_{n_j} = \underbrace{\sum_{\mu=1}^{\mu\leq i_v} r_{n_{(v,\mu)}}}_{\leq r^v} + \sum_{o\in\mathcal{O}} \underbrace{\sum_{\mu=1}^{\mu\leq i_o} r_{n_{(o,\mu)}}}_{\leq 0} \leq r^v.$$

Furthermore, we have $\rho_{n_i} \geq 0$ as the accumulated load at every point in the tour θ^* is due to (C2) greater than or equal to the accumulated load at the corresponding point in the feasible tour θ. **Q.E.D.**

Note that lemma 3 can also be applied to the GVRP-DWH.

6.4.3 REARRANGE

The REARRANGE operator rearranges locations belonging to one order to other positions within the same tour, subject to precedence constraints imposed on the GVRP.

Definition 15 *A rearrangement of an order $o \in \mathcal{O}_\theta$ is a transformation of θ to a tour θ^* with $\mathcal{O}_{\theta^*} = \mathcal{O}_\theta$ such that the relative order of all nodes not belonging to order $o \in \mathcal{O}_\theta$ remains unchanged. A rearrangement is feasible if and only if the resulting tour is feasible.*

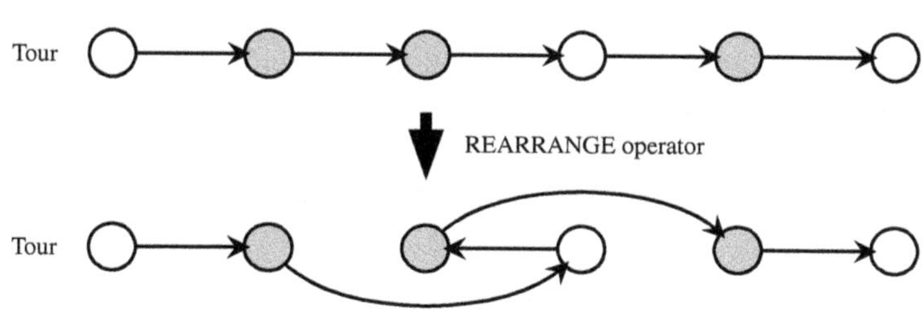

Fig. 6.5: REARRANGE

The REARRANGE operator is illustrated in figure 6.5 and can be interpreted as combination of INSERT and REMOVE operators. First, it removes an order from the tour and then, it inserts it again to the same tour. Feasibility and efficiency of a REARRANGE-move can be derived from the corresponding INSERT- and REMOVE-moves.

6.4.4 SHIFT

The SHIFT operator moves all locations belonging to an order from one tour to another, subject to compatibility and precedence constraints imposed on the GVRP.

Definition 16 *A shift of an order $o \in \mathcal{O}_{\theta_1}$ to tour θ_2 is a transformation of θ_1 and θ_2 to tours θ_1^* and θ_2^* with $\mathcal{O}_{\theta_1^*} = \mathcal{O}_{\theta_1} \setminus \{o\}$ and $\mathcal{O}_{\theta_2^*} = \mathcal{O}_{\theta_2} \cup \{o\}$ such that θ_1^* is a subsequence of θ_1 and θ_2 is a subsequence of θ_2^*. A shift is feasible if and only if the resulting tours are feasible.*

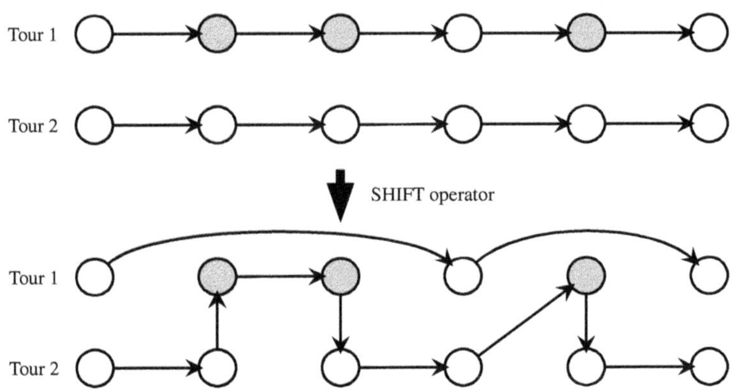

Fig. 6.6: SHIFT

The SHIFT operator is illustrated in figure 6.6 and can be interpreted as combination of INSERT and REMOVE operators. First, it removes an order from a tour and then, it inserts this order to another tour. Feasibility and efficiency of a SHIFT-move can be derived from the corresponding INSERT- and REMOVE-moves.

6.4.5 EXCHANGE

The EXCHANGE operator is a combined SHIFT operator which shifts the locations belonging to two orders in different tours to the respective other tour. Compatibility and precedence constraints imposed on the GVRP are taken into account by the EXCHANGE operator.

Definition 17 *An exchange of orders* $o_1 \in \mathcal{O}_{\theta_1}$ *and* $o_2 \in \mathcal{O}_{\theta_2}$ *is a transformation of* θ_1 *and* θ_2 *to tours* θ_1^* *and* θ_2^* *with* $\mathcal{O}_{\theta_1^*} = \mathcal{O}_{\theta_1} \setminus \{o_1\} \cup \{o_2\}$ *and* $\mathcal{O}_{\theta_2^*} = \mathcal{O}_{\theta_2} \setminus \{o_2\} \cup \{o_1\}$ *such that the relative orders of all nodes not belonging to order* $o_1 \in \mathcal{O}_{\theta_1}$ *and* $o_2 \in \mathcal{O}_{\theta_2}$ *remain unchanged. An exchange is feasible if and only if the resulting tours are feasible.*

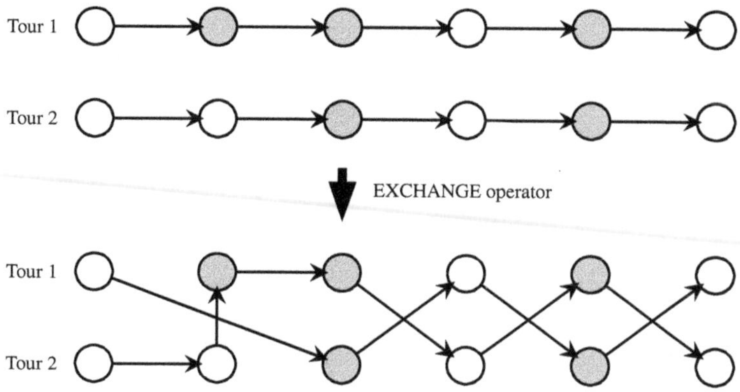

Fig. 6.7: EXCHANGE

The EXCHANGE operator is illustrated in figure 6.7 and can be interpreted as combination of INSERT and REMOVE operators. First, it removes two orders from their tours and then, it inserts these orders to the former tour of the other. Feasibility and efficiency of a EXCHANGE-move can be derived from the corresponding INSERT- and REMOVE-moves.

6.4.6 REPLACE

The REPLACE operator inserts all locations belonging to an unscheduled order to a tour, replacing all locations belonging to an order assigned to this tour. Compatibility and precedence constraints imposed on the GVRP are taken into account by the REPLACE operator.

Definition 18 *A replacement of an order* $o_2 \in \mathcal{O}_\theta$ *by an order* $o_1 \notin \mathcal{O}_\theta$ *is a transformation of* θ *to a tour* θ^* *with* $\mathcal{O}_{\theta^*} = \mathcal{O}_\theta \setminus \{o_2\} \cup \{o_1\}$ *such that the relative order of all nodes not belonging to order* $o_2 \in \mathcal{O}_\theta$ *remains unchanged. A replacement is feasible if and only if the resulting tour is feasible.*

The REPLACE operator is illustrated in figure 6.8 and can be interpreted as combination of INSERT and REMOVE operators. First, it removes an order from a tour and then, it inserts an unscheduled order to this tour. Feasibility and efficiency of a

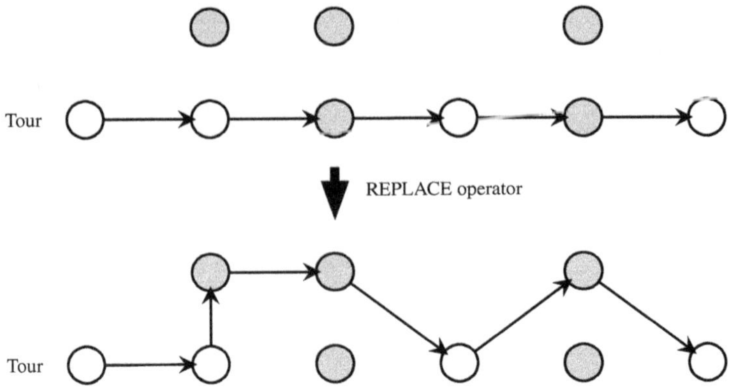

Fig. 6.8: REPLACE

REPLACE-move can be derived from the corresponding INSERT- and REMOVE-moves.

6.4.7 SHIFT-REPLACE

The SHIFT-REPLACE operator moves all locations belonging to one order from a tour to another, replacing all locations belonging to another order scheduled to this tour. Compatibility and precedence constraints imposed on the GVRP are taken into account by the SHIFT-REPLACE operator.

Definition 19 *A shift-replacement of an order* $o_2 \in \mathcal{O}_{\theta_2}$ *by an order* $o_1 \in \mathcal{O}_{\theta_1}$ *is a transformation of* θ_1 *and* θ_2 *to tours* θ_1^* *and* θ_2^* *with* $\mathcal{O}_{\theta_1^*} = \mathcal{O}_{\theta_1} \setminus \{o_1\}$ *and* $\mathcal{O}_{\theta_2^*} = \mathcal{O}_{\theta_2} \setminus \{o_2\} \cup \{o_1\}$ *such that* θ_1^* *is a subsequence of* θ_1 *and the relative order of all nodes in* θ_2 *not belonging to order* $o_2 \in \mathcal{O}_{\theta_2}$ *remains unchanged. A shift-replacement is feasible if and only if the resulting tours are feasible.*

The SHIFT-REPLACE operator is illustrated in figure 6.9 and can be interpreted as combination of INSERT and REMOVE operators. First, it removes two orders from their tours and then, it inserts one of these orders to the former tour of the other. Feasibility and efficiency of a SHIFT-REPLACE-move can be derived from the corresponding INSERT- and REMOVE-moves.

6.5 Insertion methods

This section presents two insertion methods and a basic tour improvement method. The advantage of these methods is that they can quickly improve a solution considering new transportation requests arriving dynamically.

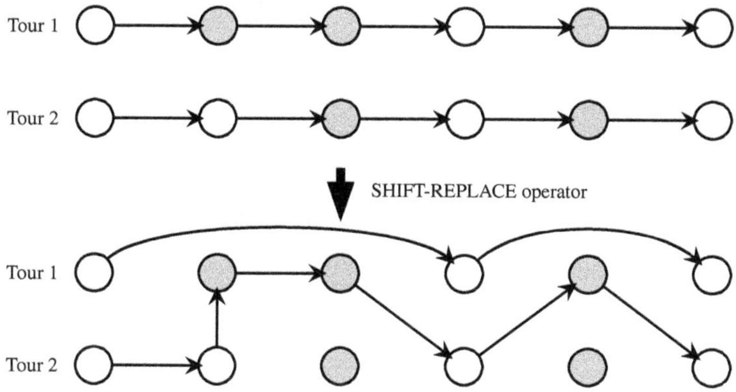

Fig. 6.9: SHIFT-REPLACE

Lemma 4 *Under assumptions (A) and either (C1) or (C2) every feasible tour can be constructed by successive feasible insertions independently of the sequence in which the orders are inserted.*

Proof: According to lemma 3 every removal of an order from a feasible tour is feasible. For every feasible removal there exists a feasible insertion undoing the removal. Therefore, every feasible tour can be constructed by successive feasible insertions and any sequence of insertions can be used to construct the tour. **Q.E.D.**

The algorithms presented in this section iteratively insert orders to tours if the INSERT-move is feasible and efficient. Thus, the objective value of the solution is steadily increased until a local optimum is found. Orders which cannot be inserted efficiently at one point in time may be inserted efficiently after other orders are inserted to the tour. Thus, in each iteration the insertion methods have to re-consider orders which could not be inserted efficiently before. The insertion methods presented in this section stop when no further efficient insertion is possible. They have the disadvantage, that under certain circumstances the revenues of orders may be very small and no single order can be efficiently inserted. In those cases inefficient insertions may need to be considered.

6.5.1 Sequential insertion

The sequential insertion method successively chooses unscheduled orders and inserts them to the tour of a vehicle. The method is illustrated in figure 6.10. First, it chooses an unscheduled order and determines all feasible insertion possibilities. If the order can be feasibly inserted, the methods chooses an efficient INSERT-move and inserts the order to the tour. If no feasible and efficient insertion possibility is found, the order remains unscheduled. If no order can be inserted efficiently, the method stops. Otherwise, it continues with the next iteration.

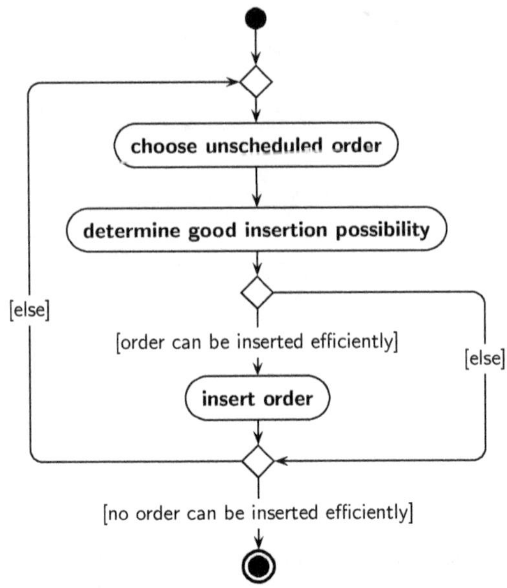

Fig. 6.10: Sequential method

Obviously, the quality of solutions generated by this method is highly dependent on the sequence in which transportation requests are chosen for insertion. Mitrović-Minić (2001) has proposed a similar insertion method for the dynamic PDP in which the set of transportation requests is sorted according to the *urgency* of the order i.e. the latest time the order has to be completed, or the *difficulty* i.e. orders which appear to be difficult to insert are chosen before others which appear easier to insert. It is argued that a transportation request is more difficult to serve, if the travel time required for visiting all locations of the transportation request is large with respect to its time windows. An analogous definition of the degree of difficulty for the GVRP would be

$$\text{difficulty} = \left(t^{\max}_{n_{(o,\lambda_o)}} - t^{\min}_{n_{(o,1)}}\right) - \min_{v \in \mathcal{V}} \sum_{i=1}^{i<\lambda_o} d^v_{n_{(o,i)} n_{(o,i+1)}}.$$

In the GVRP it is more complicated to determine a measure for *difficulty* due to the large variety of real-life requirements. Furthermore, different revenues of orders must be taken into account to prevent the method to insert orders with low profitability before highly profitable orders are considered.

6.5.2 Parallel insertion

The parallel insertion method simultaneously inserts several unscheduled orders. It is based on the auction algorithm proposed for the VRPTW by Antes and Derigs

(1995). The method is illustrated in figure 6.11 and can be described as an auctioning process or market game with proposals and acceptance.

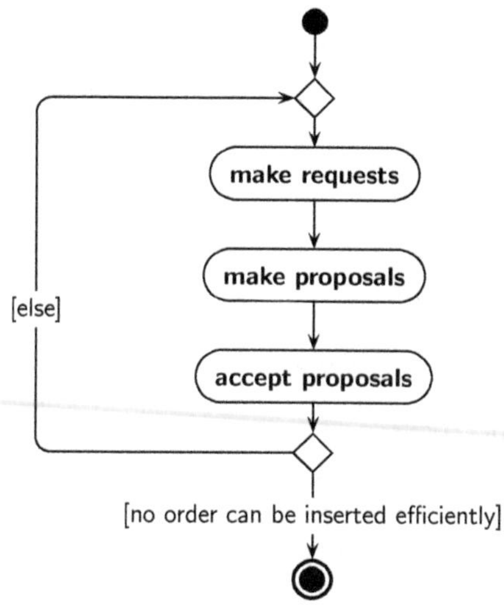

Fig. 6.11: Auction method

With this approach it is unlikely that transportation requests with low efficiency prevent highly profitable transportation requests from being inserted, as latter are preferred during the auctioning process. As a result, schedules generated from scratch are usually of much better quality than those obtained with the sequential insertion method.

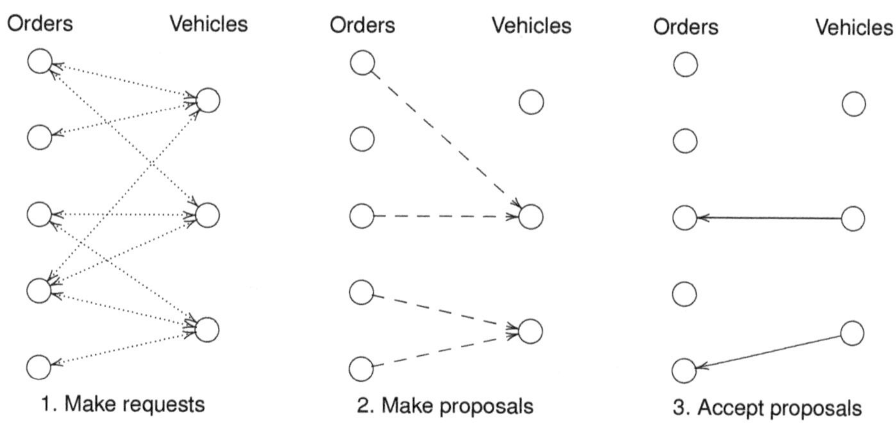

Fig. 6.12: Illustration of the auctioning process

An iteration of the auction method can be divided into three phases which are illustrated in figure 6.12. In the first phase, all unscheduled orders request and receive from each vehicle an insertion possibility and the incremental costs. An infinite cost is assumed if no feasible insertion is possible. In the second phase, each unscheduled order, which received an efficient insertion possibility, chooses a vehicle with low incremental costs and sends a proposal for insertion to this vehicle. In phase three, each vehicle which received a proposal chooses an order with high efficiency for insertion to the tour. The method stops if no order can be efficiently inserted, and continues with the next iteration otherwise.

6.5.3 Basic tour improvement

The insertion methods described above generate locally optimal solutions with respect to the INSERT operator. The solutions, however, are not necessarily locally optimal with respect to the REMOVE operator. The example illustrated in figure 6.14 shows that orders can be efficiently removed from their tour, even if each insertion was efficient at the time it has been performed. In this example each transportation request is a full-truckload pickup and delivery request. After orders 1, 4, and 3 have been efficiently inserted, order 1 can be efficiently removed from the tour. The removal not only increases the objective value of the solution, but also allows order 2 to be efficiently inserted to the tour.

Figure 6.13 illustrates a basic improvement method. The method starts by invoking an insertion method which generates a locally optimal solution with respect to the INSERT operator. Then, it iteratively removes orders from their tours if the removals are efficient. If at least one order is removed the method continues with another call of the insertion method. Otherwise, it stops with a solution that is locally optimal with respect to the INSERT operator and the REMOVE operator.

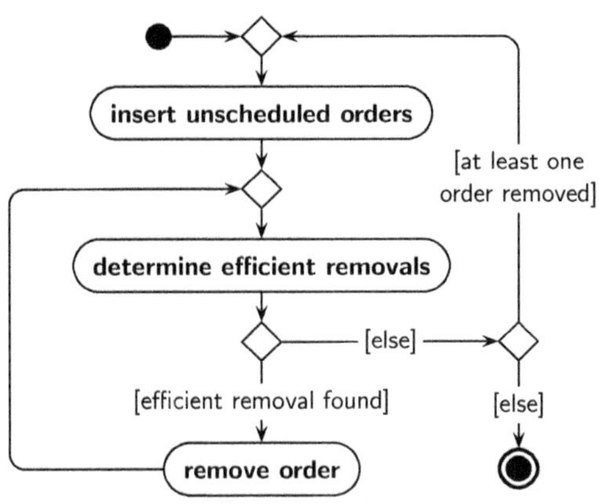

Fig. 6.13: Basic tour improvement method

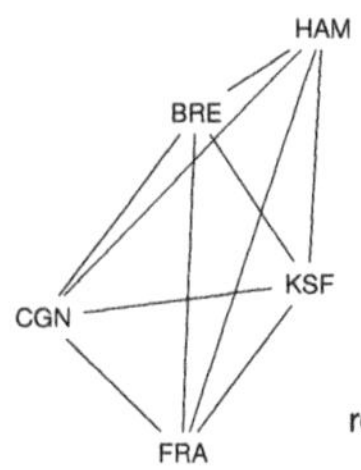

Vehicle: FRA → HAM

Order 1: FRA → KSF

Order 2: FRA → CGN

Order 3: CGN → BRE

Order 4: BRE → HAM

revenue of all orders is 300

	FRA	CGN	KSF	BRE	HAM
FRA	0	190	240	490	530
CGN	190	0	260	300	430
KSF	240	260	0	270	300
BRE	490	300	270	0	130
HAM	530	430	300	130	0

distance and cost matrix

Position	Location	Cost	Loads	Effic.
START	FRA			
END	HAM	530		-530

0. Initial tour

Position	Location	Cost	Loads	Effic.
START	FRA			
0.	**FRA**	**0**	**empty**	
1.	**KSF**	**240**	**Order 1**	**290**
END	HAM	540		-240

1. Insertion

Position	Location	Cost	Loads	Effic.
START	FRA			
0.	FRA	0	empty	
1.	KSF	240	Order 1	280
2.	**BRE**	**510**	**empty**	
3.	**HAM**	**640**	**Order 4**	**100**
END	HAM	640		-40

2. Insertion

Position	Location	Cost	Loads	Effic.
START	FRA			
0.	FRA	0	empty	
1.	KSF	240	Order 1	-10
2.	**CGN**	**500**	**empty**	
3.	**BRE**	**800**	**Order 3**	**10**
4.	BRE	800	empty	
5.	HAM	930	Order 4	300
END	HAM	930		-30

3. Insertion

Position	Location	Cost	Loads	Effic.
START	FRA			
2.	CGN	190	empty	
3.	BRE	490	Order 3	300
4.	BRE	490	empty	
5.	HAM	620	Order 4	300
END	HAM	620		-20

4. Removal

Position	Location	Cost	Loads	Effic.
START	FRA			
0.	**FRA**	**0**	**empty**	
1.	**CGN**	**90**	**Order 2**	**300**
2.	CGN	190	empty	
3.	BRE	490	Order 3	300
4.	BRE	490	empty	
5.	HAM	620	Order 4	300
END	HAM	620		280

5. Insertion

Fig. 6.14: Although each insertion is efficient the solution can be improved by an efficient removal

6.6 Reduced Variable Neighbourhood Search

This section introduces an algorithm presented by Goel and Gruhn (2007) following the Reduced Variable Neighbourhood Search (RVNS) scheme described in Hansen and Mladenović (2003). As illustrated in figure 6.15, a locally optimal solution with respect to one neighbourhood structure is not necessarily locally optimal for another neighbourhood structure.

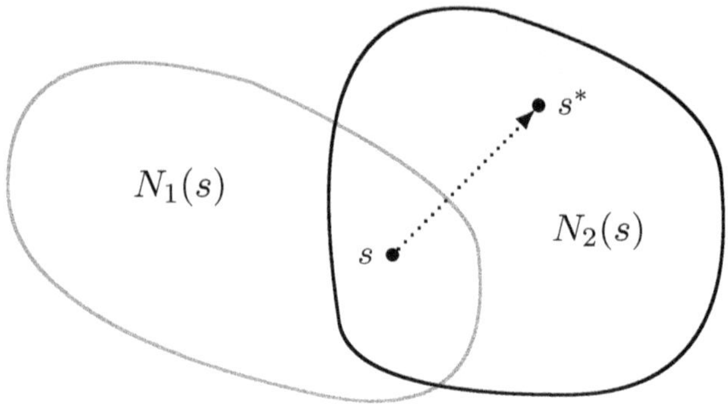

Fig. 6.15: s is a locally optimal solution in neighbourhood $N_1(s)$ but not in neighbourhood $N_2(s)$. Thus, the current solution can be improved by replacing s with the locally optimal solution s^* in $N_2(s)$.

The RVNS algorithm exploits this observation by changing the neighbourhood structure during the search. In each iteration the RVNS explores one of the following neighbourhoods in order to find an improving solution:

Neighbourhood 1: The first neighbourhood is defined by the feasible solutions that can be found by choosing an unscheduled order and applying the INSERT operator.

Neighbourhood 2: The second neighbourhood is defined by the feasible solutions that can be found by choosing a scheduled order and applying the REMOVE operator, the REARRANGE operator, or the SHIFT operator.

Neighbourhood 3: The third neighbourhood is defined by the feasible solutions that can be found by choosing an unscheduled and a scheduled order and applying the REPLACE operator.

Neighbourhood 4: The fourth neighbourhood is defined by the feasible solutions that can be found by choosing two orders scheduled to different tours and applying the SHIFT-REPLACE operator or the EXCHANGE operator.

Given these neighbourhood structures, the RVNS algorithm can be outlined as illustrated in figure 6.16.

The algorithm starts with an initial solution s. Until a stopping criterion is met, e.g. the maximum computing time, the RVNS algorithm repeats the following steps. First, the next neighbourhood $N_k(s)$ to be considered is chosen randomly. Then, a new solution $s^* \in N_k(s)$ is generated. The new found solution s^* is accepted as the next current solution if the objective value is improved.

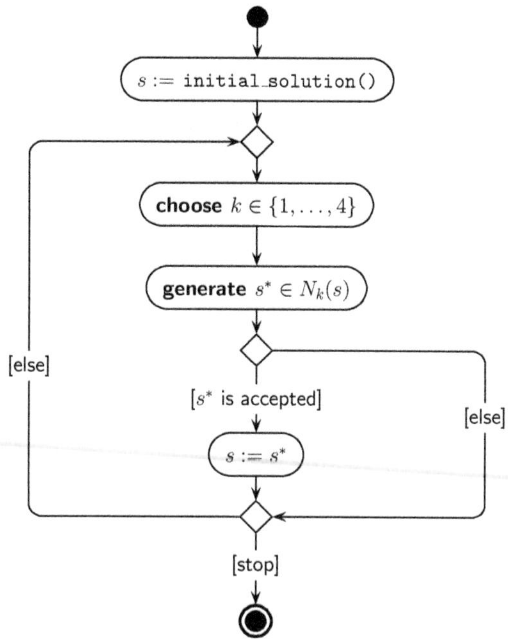

Fig. 6.16: Reduced Variable Neighbourhood Search

6.7 Large Neighbourhood Search

This section describes the Large Neighbourhood Search (LNS) algorithms presented by Goel and Gruhn (2007). LNS allows to escape from locally optimal solutions of poor quality as illustrated in figure 6.17. In each iteration some of the orders are

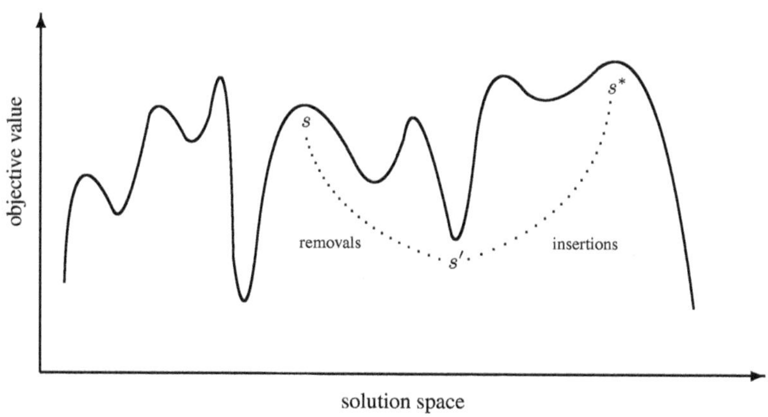

Fig. 6.17: Pictorial representation of a Large Neighbourhood Search move.

removed from their tours in the current solution s, leading to an interim solution s' with presumably worse objective function value. Then, unscheduled orders are re-inserted until another locally optimal solution s^* is found.

Let s denote a feasible solution and let $|\mathcal{O}_s|$ denote the number of transportation requests which are assigned to the tour of some vehicle. For each solution s let $N_k(s)$ denote the kth neighbourhood of s which is defined by removing k transportation requests from their tours. Then, the LNS algorithm can be outlined as illustrated in figure 6.18.

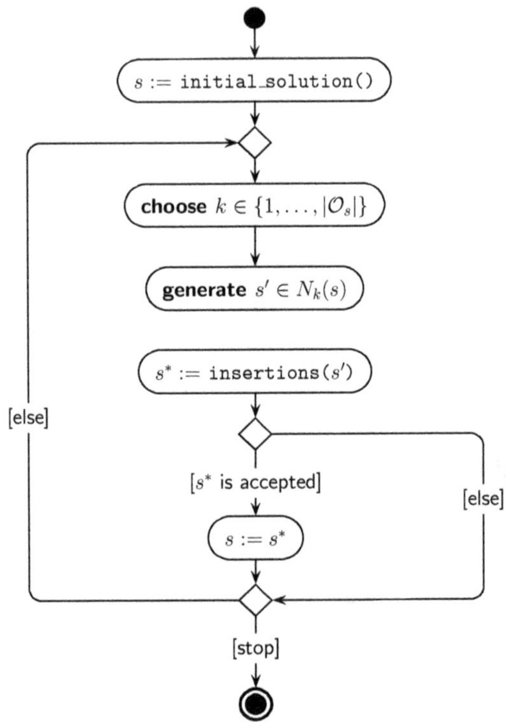

Fig. 6.18: Large Neighbourhood Search

The algorithm starts with an initial solution s. In each iteration the number k of transportation requests to be removed is chosen. Then, an interim solution s' is generated by removing k transportation requests from the tours they are currently assigned to. A new solution s^* is generated using one of the insertion methods presented in section 6.5. The new solution is accepted as the next current solution if the objective value is improved. If no stopping condition is met, the algorithm continues with the next iteration.

The probability that a better solution can be found is strongly connected with the choice of the neighbourhood $N_k(s)$. If k is too small, the solutions which can be found by an LNS move will be very similar to the current solution. If k is too large, the insertion method may need too much time and the new solution found may not be much better than a solution generated from scratch.

6.7.1 Removals

The goal of removing transportation requests is to generate an auspicious interim solution such that the insertion method can find a new solution with better quality. A simple approach is to randomly remove k transportation requests from the tours as illustrated in figure 6.19.

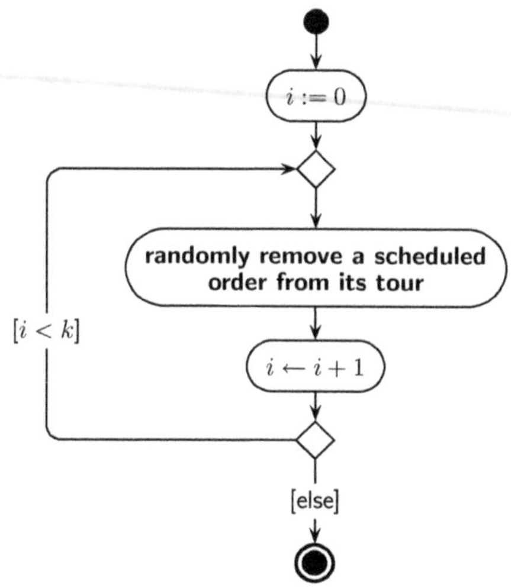

Fig. 6.19: Random removals

If transportation requests are removed randomly, some of them may not be *related* to each other in any way. Hence, the re-insertions of these transportation requests are independent of another and the same effect can be achieved by removing less transportation requests and performing the re-insertions sequentially.

For the VRP, Shaw (1997) propose a relatedness measure based on geographical closeness of customer locations. A concept similar to geographical closeness in the VRP, however, does not exist if transportation requests can have multiple pickup, delivery and/or service locations. If geographical closeness cannot be used, the question is how to define a relatedness measure for the GVRP. We want to increase the probability that a transportation request which is removed from the tour of a vehicle allows another transportation request to take its "place".

Goel and Gruhn (2007) proposed a tour dependent relatedness measure to achieve this goal. An order assigned to the tour of a vehicle which is not suited for an unscheduled order is not regarded related to the latter. An order assigned to the tour

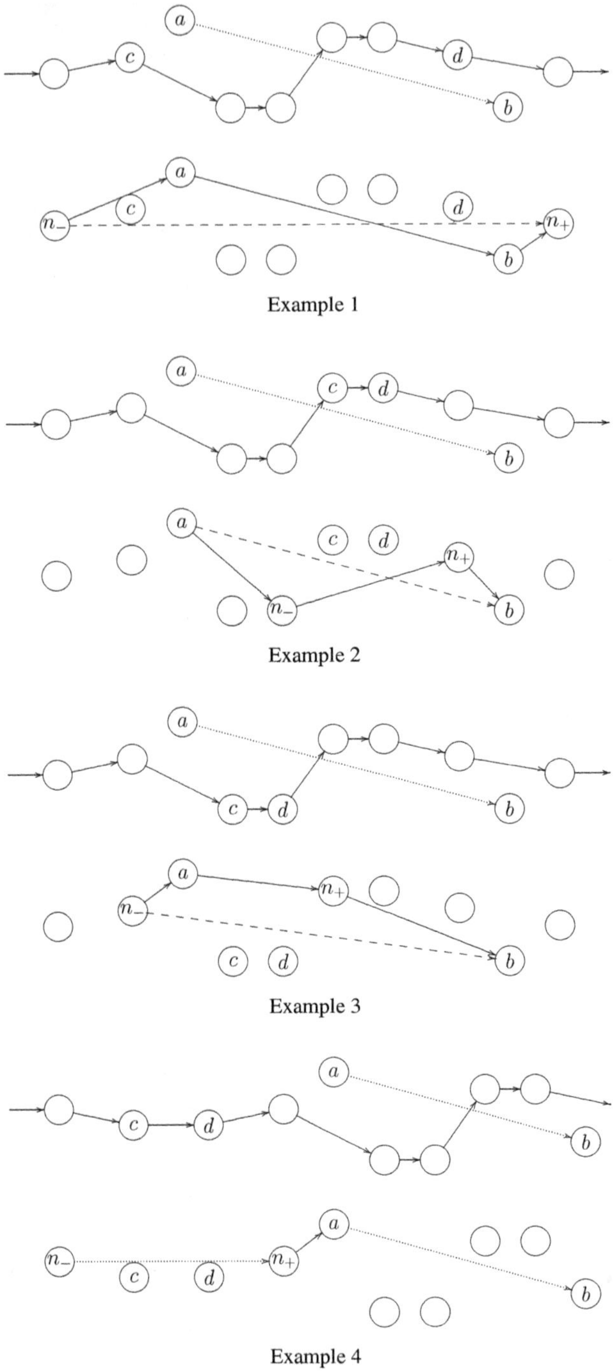

Fig. 6.20: Determination of relatedness values

of a vehicle which is suited for an unscheduled order is regarded related if the unscheduled order o "fits" to the part of the tour currently occupied by the former. The relatedness measure is determined as illustrated in the examples in figure 6.20. In the illustration ⓐ represents node $n_{(o,1)}$ and ⓑ represents node $n_{(o,\lambda_o)}$. To determine the relatedness value of a scheduled order (represented by nodes ⓒ and ⓓ in the illustration), the scheduled order is not regarded itself, but its preceding and succeeding point in the tour. Let n_- and n_+ denote the predecessor and successor of the scheduled order. Now all sequences $\theta = (n_1, \ldots, n_{\lambda_o+2})$ containing the subsequences (n_-, n_+) and $(n_{(o,1)}, \ldots, n_{(o,\lambda_o)})$ are determined, for which arrival times $t_{n_1}, \ldots, t_{n_{\lambda_o+2}}$ exist with

$$t_{n_i} + d^v_{n_i n_{i+1}} \leq t_{n_{i+1}} \text{ for all } 1 \leq i < \lambda_o + 2$$

and

$$t^{\min}_{n_i} \leq t_{n_i} \leq t^{\max}_{n_i} \text{ for all } 1 \leq i \leq \lambda_o + 2.$$

Obviously, there may be various such sequences. Among these, let us consider the sequence $\theta = (n_1, \ldots, n_{\lambda_o+2})$ with least costs

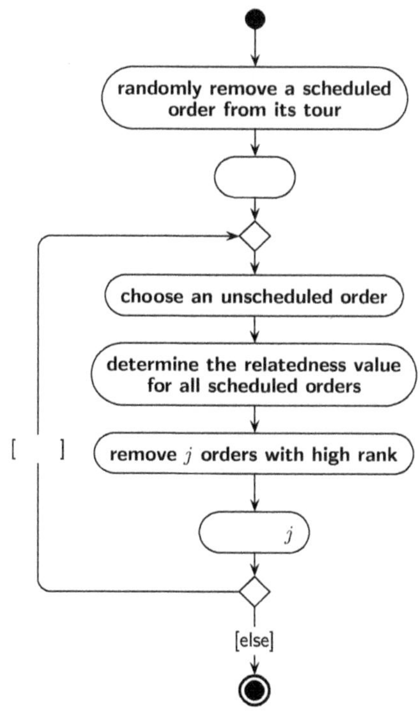

Fig. 6.21: Removal of transportation requests using the relatedness measure

$$c_\theta := \sum_{i=1}^{i<\lambda_o+2} c^v_{n_i n_{i+1}}.$$

If, as in examples 1-3 of figure 6.20, n_+ is visited after $n_{(o,1)}$ and n_- is visited before $n_{(o,\lambda_o)}$ the relatedness value is

$$\text{relatedness value} := c_\theta - c^v_{n_1 n_{\lambda_o+2}}.$$

Otherwise, n_- is visited after $n_{(o,\lambda_o)}$ or n_+ is visited before $n_{(o,1)}$, as shown in example 4 of figure 6.20. If n_+ is visited before $n_{(o,1)}$, the scheduled order is not regarded related to the unscheduled order if the removal from the tour would not allow n_+ to be visited earlier. Analogously, if n_- is visited after $n_{(o,\lambda_o)}$, the scheduled order is not regarded related to the unscheduled order if the removal from the tour would not allow n_- to be visited later. Otherwise, the relatedness value is the cost for travelling from n_+ to $n_{(o,1)}$ or from $n_{(o,\lambda_o)}$ to n_-.

A small relatedness value indicates that the unscheduled order would be an auspicious candidate for insertion if the considered scheduled order was removed from the tour. Using this relatedness measure we can remove transportation requests as illustrated in figure 6.21. First, a randomly chosen transportation request is removed from its tour. Then, an unscheduled transportation request is chosen randomly, and for all scheduled transportation requests the relatedness value is determined. The related transportation requests are ranked according to their relatedness value and some of them with high rank are chosen to be removed.

6.7.2 Re-insertion

After orders have been removed from their tours, the LNS algorithm uses either the sequential or the parallel insertion method, and the basic tour improvement procedure presented in section 6.5. Within the insertion methods, the INSERT-moves are evaluated locally but their effect on the global solution quality is not known. If only the best INSERT-moves are considered only a part of the solution space is explored and the resulting schedule may be of poor quality.

As the example in figure 6.22 illustrates, it may be better to occasionally chose INSERT-moves which are not locally optimal. In the example it is assumed that the two orders must not be served simultaneously. Each order can be inserted to the tour of the vehicle by INSERT-moves with different efficiency. The least cost tour covering both orders, however, is only found if the second best INSERT-move for one of the orders is chosen. As we can see in this example, choosing the second best alternative can result in finding the optimal solution, while always choosing the best alternative results in a solution of much lower quality.

To increase diversification, i.e. to allow the exploration of a larger part of the solution space and to avoid getting repeatedly trapped in a local optimum of poor quality, the degree of determinism can be decreased. This can be done by ranking the INSERT-moves according to their efficiency. Within this ranked list an INSERT-move with high rank is chosen randomly. Let $q \in [0, 1)$ be a random number and

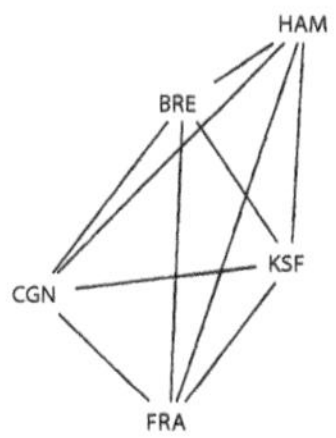

Vehicle: KSF → CGN → HAM

Order 1: KSF → BRE

Order 2: FRA → HAM

	FRA	CGN	KSF	BRE	HAM
FRA	0	190	240	490	530
CGN	190	0	260	300	430
KSF	240	260	0	270	300
BRE	490	300	270	0	130
HAM	530	430	300	130	0

distance and cost matrix

Tours found by applying the most efficient INSER T-move for each order:

Tours found by applying the second best INSER T-move for each order:

Optimal tour covering both orders:

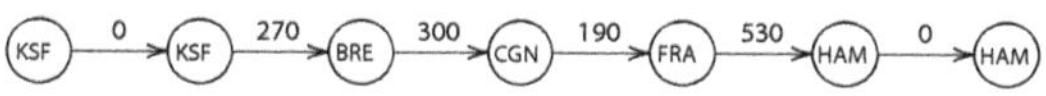

Fig. 6.22: Optimal solution may not be found if only locally optimal moves are made

let $\alpha \geq 1$ be a positive number. Then the INSERT-move with rank $q^\alpha * |M| + 1$ is chosen, where M denotes the set of all feasible INSERT-moves. If $\alpha = \infty$ only the most efficient INSERT-move is chosen. If $\alpha = 1$ the choice is only based on the value of the random value q. The value of α can be adjusted dynamically during computation.

6.8 Evaluation and case study

This chapter presented heuristics for solving the dynamic GVRP and GVRP-DWH. In order to evaluate the algorithms presented, test cases have been generated incorporating most of the complexities found in the problem, the motor carrier of the case study introduced in section 3.6 has to solve. In the case study the carrier has to transport shipments between European airports. A frequency distribution of pickups and deliveries at these airports is assumed as illustrated in figure 6.23, in which the frequency of pickups and deliveries at the airports is indicated by the size of the circle. Most of the shipments have to be picked up or delivered to an airport in the region between Paris, Düsseldorf, and Frankfurt. Some shipments, however, have very remote origins or destinations, for example Florence, Dublin, Gothenburg, and Helsinki.

A heterogeneous vehicle fleet has been generated where some of the vehicles have refrigerated cargo bodies and some are manned by two drivers. All vehicles have a capacity of 4D, i.e. at most four ULD of type D can be loaded. It is assumed

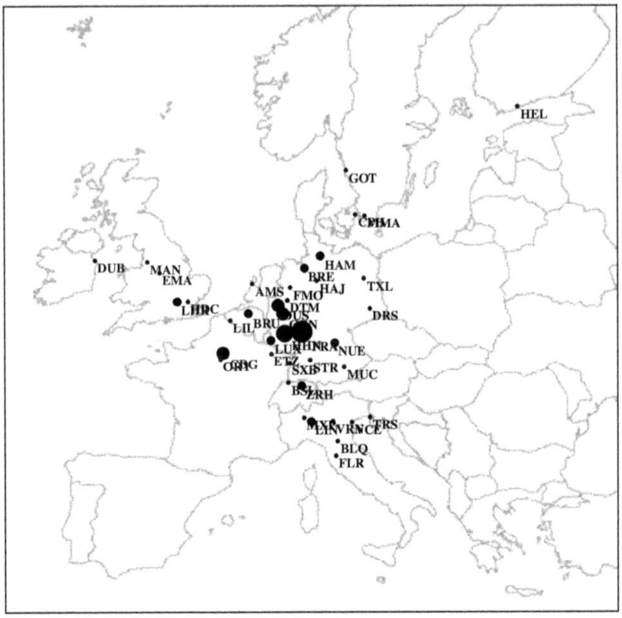

Fig. 6.23: Distribution of pickup and delivery locations

that all vehicles are en-route when planning starts and each vehicle becomes available at one of the airports during some time of the day. All vehicles eventually have to return to the depot in Frankfurt.

Three types of orders were generated in a way that the frequency distribution illustrated in figure 6.23 is achieved:

- requests for transporting two or four ULD from a pickup location to a delivery location,
- requests for picking up four ULD at their origin and delivering two ULD to one destination and two ULD to another, and
- requests for picking up two ULD at one origin and two ULD at another, and delivering all ULD to the same destination.

Some of the transportation requests require a vehicle with refrigerated cargo body, some require a vehicle manned by two drivers.

Travel distances are based on the direct distances between the airports. In order to consider the average deviation occurring in road transport they are multiplied by 1.3. Travel costs are proportional to the travel distance. Vehicles with refrigerated cargo bodies and vehicles manned by two drivers are more expensive than vehicles with standard cargo bodies and those manned by one driver. For all orders it is assumed that external carriers can be employed at double the costs of the cheapest vehicle capable of transporting the shipments, i.e. the cost for the transport plus the costs for an empty return trip. In the beginning only some of the orders are known to the carrier and every hour in the simulation scenario new orders arrive. For all orders the length of the time windows at each location is set to the same value.

Two types of test cases were generated: test cases for the GVRP and test cases for the GVRP-DWH. In the GVRP test cases an average speed of 65 km/h is assumed for vehicles manned by one driver, and 70 km/h for vehicles manned by two drivers. In the GVRP-DWH test cases an average speed of 75 km/h is assumed for all vehicles.

Computational experiments were performed on a personal computer with Intel Pentium 4 processor with 3.00 GHz. In the simulation of 10 hours of dynamic planning the algorithms were only allowed 60 seconds of computing time per timestep (representing one hour in the simulation scenario). At each timestep all previously unscheduled transportation requests are assumed to be subcontracted by external carriers and removed from the model. Afterwards, new transportation requests are inserted to the tours by the auction method presented in section 6.5.

Table 6.1 and table 6.2 show the results of the computational experiments. The following notation is used in the tables:

- $|\mathcal{V}|$ the number of vehicles
- $|\mathcal{O}_0|$ the number of orders known at the beginning of the simulation
- $|\mathcal{O}_t|$ the numbers of orders that arrive dynamically at timestep t
- τ the length of the time windows (in hours)
- AVG(f) the average value of the objective function
- AVG(Δ) the mean absolute deviation of the objective function value
- RVNS the Reduced Variable Neighbourhood Search method
- LNSU-S the LNS method with random (unrelated) removals using the sequential method for re-insertion
- LNSU-A the LNS method with random (unrelated) removals using the auction method for re-insertion
- LNSR-S the LNS method using the relatedness measure for removing transportation requests and the sequential method for re-insertion
- LNSR-A the LNS method using the relatedness measure for removing transportation requests and the auction method for re-insertion

In each iteration of the LNS algorithms the number of transportation requests to be removed was chosen randomly such that $k \in [2, 30]$. As we can see in table 6.1 neither RVNS, nor LNSU-S or LNSU-A clearly dominate each other for the GVRP test cases. While LNSU-S and LNSU-A seam to perform better for small problems, RVNS seems to perform better for the large instances. The use of the relatedness criterion significantly improves the performance of the LNS method and LNSR-S and LNSR-A clearly outperform LNSU-S and LNSU-A. LNSR-A produces the best average results in almost all cases. For the large instances, however, RVNS is competitive.

In table 6.2 we can see that the use of the relatedness criterion is counter productive if drivers' working hours are considered and LNSR-S and LNSR-A are clearly outperformed by LNSU-S and LNSU-A. Furthermore, RVNS does not perform well for the GVRP-DWH instances. The LNS algorithms using the auction method outperform those using the sequential insertion method. LNSU-A produces the best average results in almost all cases.

For both problem types response times of all algorithms were below a few seconds and average response times were mostly below one second. Due to the smaller size of the neighbourhoods to be explored, the RVNS algorithm has much smaller response times than the LNS algorithms. It appears that the best performance of the LNS approach can generally be achieved by using the auction method for reinsertion. The performance of the relatedness measure, however, seems to be problem dependent and different relatedness measures may have to be developed for problem instances with different characteristics.

As mentioned above, the size of the neighbourhoods used by the RVNS method is smaller than the average neighbourhood size used by the LNS approach. This is probably the reason why RVNS does not perform well for the GVRP-DWH test cases, in which the RVNS struggles in moving from the current solution to a neighbouring solution which significantly improves the objective function value. However, the exploration of the smaller neighbourhoods is more intense and thus, the RVNS method performs well for the large GVRP test cases. It seems that the higher diversification of the LNS approach is not as effective for these large instances. Of course, the intensification of the search can be increased if the number k of transportation requests to be removed in an iteration of LNS is reduced. Note that for $k \leq 2$ the neighbourhood structures of LNS and RVNS are very similar as all neighbourhood operators used by the RVNS method can be interpreted as a combination of at most two removals and subsequent insertions.

As the GVRP generalises the classical models, the algorithms presented in this chapter may also be used for instances of these problems. In order to tackle such instances, the revenues p_o of all orders $o \in \mathcal{O}$ are set to very large values, such that the heuristics are encouraged to only produce solutions in which all orders are served. Therefore, the RVNS method will not be very effective as only the neighbourhood operators REARRANGE, SHIFT, and EXCHANGE allow to move from one solution in which all transportation requests are served to another solution in which all transportation requests are served. The effectivity of the LNS approach which was originally developed for the VRPTW, however, will not be seriously affected. Note that highly specialised algorithms developed for the classical vehicle routing problems can exploit problem-specific knowledge and should have better performance than any method developed for the GVRP or GVRP-DWH. The goal of this work, however, was to develop algorithms that can be used for problems considering a variety of practical complexities that are not considered by the classical models.

Obviously, it is impossible to generalise the presented results to any possible case of application. In particular, if the algorithms are used in an interactive real-time de-

Problem					RVNS		LNSU-S		LNSU-A		LNSR-S		LNSR-A	
No.	$\|\mathcal{V}\|$	$\|\mathcal{O}_0\|$	$\|\mathcal{O}_t\|$	τ	AVG(f)	AVG(Δ)	AVG(f)	AVG(Δ)	AVG(f)	AVG(Δ)	AVG(f)	AVG(Δ)	AVG(f)	AVG(Δ)
p1	100	300	20	2	181038.00	1906.25	184029.14	919.38	186550.44	464.69	186606.94	4053.05	193658.71	1106.38
p2	100	300	20	2	197575.33	1155.46	208776.15	1954.39	205134.57	2040.47	212594.54	2420.89	219957.12	690.86
p3	100	300	20	2	188633.12	1171.72	196356.79	1062.86	196017.53	2103.10	197014.18	865.80	198857.20	1418.95
p4	100	300	20	2	185183.01	1303.51	192348.58	2804.70	191217.30	3084.49	195481.51	1054.85	196001.81	1909.06
p5	100	300	20	12	289579.38	1617.78	292741.27	2421.49	297616.97	1454.91	296927.72	1695.53	299628.28	1679.21
p6	100	300	20	12	273997.70	1483.08	277866.03	3102.23	278874.72	2281.10	282781.13	5314.64	289026.76	1087.93
p7	100	300	20	12	262702.54	1145.46	269749.28	1256.22	268784.42	1015.83	271448.41	1523.97	272337.89	882.10
p8	100	300	20	12	271150.15	1330.98	280591.50	3481.43	281425.79	800.88	283126.79	2591.13	287839.02	2451.13
p9	250	750	50	2	505072.76	4998.03	507704.97	2862.47	506442.17	1892.44	523771.22	4695.66	527543.21	4929.89
p10	250	750	50	2	487759.72	1998.89	486023.13	2332.94	488689.09	1529.98	492162.40	4179.56	497543.61	2786.64
p11	250	750	50	2	502367.70	1979.73	505533.84	819.25	498615.52	2548.65	520161.15	1858.15	511879.04	3378.25
p12	250	750	50	2	526699.71	6040.94	521351.02	3111.42	518665.58	2871.04	521442.66	2978.28	532709.90	4177.82
p13	250	750	50	12	713547.18	4179.45	703292.48	5994.75	717727.04	2146.32	717338.79	6886.78	738878.17	4699.38
p14	250	750	50	12	722309.20	2264.87	698004.65	2183.89	712107.92	4972.12	727244.41	6379.60	721670.85	3003.67
p15	250	750	50	12	689475.10	2391.16	677776.26	6827.64	679460.74	1732.82	682308.68	4361.44	692698.84	3064.41
p16	250	750	50	12	702658.55	6867.02	679445.38	4452.84	682183.45	2705.21	690590.02	4428.49	692122.23	1294.61
p17	500	1500	100	2	1083725.74	4658.40	1067629.88	3167.80	1069182.97	2929.16	1114487.76	6039.94	1106940.06	9608.54
p18	500	1500	100	2	1062635.10	8941.89	1043420.53	2117.96	1065859.76	7547.59	1061797.30	9131.62	1089812.91	2892.04
p19	500	1500	100	2	1022168.34	6425.85	997030.59	172.18	1022978.35	3805.23	1042586.13	1380.33	1032741.35	4857.31
p20	500	1500	100	2	1059648.63	7331.57	1023884.03	5823.38	1026574.20	4489.41	1054628.70	11642.70	1065159.95	12771.29
p21	500	1500	100	12	1395467.83	8079.59	1389883.00	4841.83	1377983.56	11195.43	1398659.64	13204.95	1399397.98	5729.63
p22	500	1500	100	12	1430855.62	5899.86	1406401.24	5769.99	1400686.31	4651.54	1437587.27	5940.28	1439543.12	8811.84
p23	500	1500	100	12	1432836.25	4466.95	1415654.38	3980.92	1415294.39	7966.82	1433025.43	5827.46	1430347.23	6648.53
p24	500	1500	100	12	1431020.98	3019.13	1418357.20	14196.09	1434010.56	8135.41	1435894.40	12400.59	1438788.69	7541.93

Table 6.1: Results of the GVRP test cases

Problem					RVNS		LNSU-S		LNSU-A		LNSR-S		LNSR-A	
No.	$\|\mathcal{V}\|$	$\|\mathcal{O}_0\|$	$\|\mathcal{O}_t\|$	τ	AVG(f)	AVG(Δ)	AVG(f)	AVG(Δ)	AVG(f)	AVG(Δ)	AVG(f)	AVG(Δ)	AVG(f)	AVG(Δ)
p1	100	300	20	2	91742.14	694.17	135753.39	2220.90	137381.34	2865.52	132855.82	1190.80	136987.49	3079.05
p2	100	300	20	2	86726.90	470.92	122595.31	1554.23	124830.97	584.05	119429.16	577.36	120658.17	2015.57
p3	100	300	20	2	98827.95	840.22	134826.81	1132.56	135249.10	674.58	133711.73	2433.00	135599.00	3055.56
p4	100	300	20	2	103292.76	308.12	144491.56	2142.04	148652.44	1992.39	143628.57	2918.27	148022.89	1307.04
p5	100	300	20	12	155682.01	1071.18	201620.45	4134.05	202496.91	1468.62	197967.20	1442.75	190191.59	2814.60
p6	100	300	20	12	162604.68	635.21	205243.72	1434.00	203922.78	4081.44	205687.85	2181.21	202654.39	2894.30
p7	100	300	20	12	157589.90	1367.47	204107.63	1871.90	204054.91	3319.60	196241.52	1235.23	199470.41	2026.28
p8	100	300	20	12	164516.86	1894.68	214839.57	2708.35	215337.51	2429.59	200723.09	5109.71	205747.62	4670.25
p9	250	750	50	2	303372.62	1586.12	365870.23	4497.96	373465.09	5445.90	349533.80	5222.73	362631.85	5545.89
p10	250	750	50	2	297855.81	911.99	363712.55	5844.97	371404.20	2991.09	342659.65	4545.80	352835.53	4517.11
p11	250	750	50	2	279893.95	1113.82	348750.49	4922.66	361869.71	2308.78	332585.47	4194.33	341481.13	1287.20
p12	250	750	50	2	286444.81	1482.61	343209.09	5460.09	359553.26	6354.11	330153.84	2762.33	346528.34	4186.61
p13	250	750	50	12	484703.12	758.67	546420.44	2459.48	563105.14	4622.74	520414.66	2685.25	527126.12	674.25
p14	250	750	50	12	457151.68	837.54	514783.49	7184.19	518417.30	4000.02	483592.14	4082.77	494160.21	2598.75
p15	250	750	50	12	456574.40	1088.74	514794.36	3092.92	532497.43	2976.30	490795.22	3300.66	496583.22	3520.88
p16	250	750	50	12	477127.29	1027.74	536386.51	7208.88	542029.30	7128.13	497219.72	10419.89	504757.13	1224.98
p17	500	1500	100	2	604958.04	803.52	655684.31	2288.21	670920.94	13598.94	612297.30	4091.49	634979.30	1066.65
p18	500	1500	100	2	630450.01	1148.93	686546.44	3593.01	691416.96	1400.10	653492.57	9819.57	676345.45	4323.51
p19	500	1500	100	2	629969.09	2712.67	697009.79	41.52	713357.25	6686.20	663912.50	3292.20	675803.63	4389.13
p20	500	1500	100	2	591035.71	1963.82	649970.33	6139.97	665278.51	2206.50	610786.76	1639.06	621275.35	3916.04
p21	500	1500	100	12	930469.72	1310.01	979649.71	6222.09	999714.67	201.88	923273.29	1831.76	945747.96	3580.34
p22	500	1500	100	12	949612.72	1292.05	976795.07	8865.35	986565.17	6189.27	957559.09	3057.07	964971.97	7040.81
p23	500	1500	100	12	925413.76	3168.22	994303.54	7821.58	1004896.40	4767.71	931428.54	2488.93	937159.35	168.28
p24	500	1500	100	12	950116.36	1186.17	1006517.33	2349.26	1039078.49	4496.73	962854.01	2602.09	992634.67	3203.70

Table 6.2: Results of the GVRP-DWH test cases

cision support system as described in section 4.4.3. The interaction with dispatchers may lead to incomparable results as discussed in section 6.2.6. Therefore, an evaluation of an interactive real-time decision support system could only be realised in extensive field studies.

7

Conclusions

7.1 Summary

This work investigates commercial vehicle operations and how fleet telematics can be used in order to improve real-time management and planning.

Chapter 2 gave an introduction into telematics and its main enabling technologies concerned with road freight transport: wireless communication, positioning systems, and Geographical Information Systems. Transport telematics applications have been surveyed focusing on those which are of particular interest to motor carriers.

Chapter 3 gave an introduction into commercial vehicle operations and described to most important influences on the development of road freight transport: globalisation, liberalisation, deregulation, supply chain management, and just-in-time practices. The fundamentals of road freight transport have been introduced and strategic, tactical, operational, and real-time management levels have been examined. Operational and real-time tasks a motor carrier has to face in order and fleet management were discussed in detail.

Management information systems used by motor carriers to perform their tasks at the operational and real-time management level have been examined in chapter 4. First, a typical management information system without any telematics functionality was described. It was discussed how the carrier's tasks can be fulfilled if the entire communication between drivers and dispatchers is realised by mobile telephones. Potentials for improvement were identified and classified. It was shown how commercial off-the-shelf fleet telematics systems can be integrated into a typical legacy information system without telematics functionality. A Messaging & Fleet Monitoring System (MFMS) was presented that supports the communication between drivers and dispatchers and automatically analyses messages received in the dispatching office. The MFMS automatically identifies actual data, compares it with planned data, identifies discrepancies between actual and planned data, and revises the planned data according to the current state of the transportation system. A Dynamic Planning System (DPS) which can provide real-time decision support has been presented. The DPS continuously optimises the current schedule considering all changes made by dispatchers and MFMS.

In chapter 5 classical models for routing a fleet of commercial vehicles have been surveyed. The General Vehicle Routing Problem (GVRP) was introduced, which is a generalisation of the classical models and capable of considering various real-life requirements such as load acceptance and employment of external carriers, time window restrictions, multiple pickup and/or delivery locations, multi-dimensional resource requirements, and a heterogeneous vehicle fleet. Furthermore, the General Vehicle Routing Problem with Drivers' Working Hours (GVRP-DWH) was introduced which considers regulations for drivers' working hours in the European Union.

Eventually, chapter 6 discussed differences between static and dynamic planning. Solution approaches capable of handling various real-life requirements in dynamic planning were surveyed. Chapter 6 presented several neighbourhood operators for the GVRP and GVRP-DWH. Two insertion methods, a Reduced Variable Neighbourhood Search algorithm, and several variants of Large Neighbourhood Search algorithms for the dynamic GVRP and GVRP-DWH were presented. All algorithms presented are characterised by very fast response times and can be used within the Dynamic Planning System. Eventually, computational experiments were performed on benchmark problems created for the GVRP and GVRP-DWH.

7.2 Scientific contributions

This work contributes to transportation research in several ways. It identifies and classifies potentials of fleet telematics for improving the efficiency of commercial vehicle operations. It shows how commercial off-the-shelf fleet telematics systems can be integrated into a typical legacy information system without telematics functionality. A Messaging & Fleet Monitoring System is presented that retrieves messages from a commercial off-the-shelf fleet telematics system, and planned data from the legacy information system. It is shown how messages are automatically analysed in order to identify actual data, to compare it with planned data, to identify discrepancies between actual and planned data, and to revise planned data according to the actual state of the transportation system.

It is shown how computer-based real-time decision support can be provided by integrating a Dynamic Planning System into the telematics-enabled information system. The real-time decision support system achieves its strength from several specialised actors with different problem knowledge and solution techniques: dispatchers, Messaging & Fleet Monitoring System, and Dynamic Planning System. These actors collaboratively and concurrently modify the schedule. An optimistic locking scheme has been presented which guarantees that data integrity is maintained.

In real-life applications sophisticated models are equally important as good techniques for finding solutions to the problem. Even if a (globally) optimal solution could be found for a simplified problem, this solution would not be a big help, if only half of it can be applied due to real-life requirements not considered by the model. On the other hand, solutions obtained for a model that considers most of the real-life requirements, are beneficial even if the solution is not (globally) optimal. In chapter 5 the General Vehicle Routing Problem (GVRP) is introduced which gives a unifying formulation capable of handling various real-life requirements, which up to now have only been treated independently. The GVRP narrows the gap between what real-life problems require and what is considered by the model. In most of the vehicle routing literature drivers' working hours are totally ignored. This work showed how regulations regarding drivers' working hours in the European Union can be considered in vehicle routing and scheduling and introduced the General Vehicle Routing Problem with Drivers' Working Hours (GVRP-DWH).

Chapter 6 presented several neighbourhood operators for the GVRP and GVRP-DWH. Two insertion methods, a Reduced Variable Neighbourhood Search algorithm, and several variants of Large Neighbourhood Search algorithms for the dynamic GVRP and GVRP-DWH are presented. All of these methods are characterised by very fast response times and can be used within the Dynamic Planning System.

7.3 Future research

This work has shown how fleet telematics can be used to improve monitoring, control and planning of transportation processes. Future research should be conducted for providing tracking & tracing functionalities across inter-organisational boundaries and throughout the entire transport chain.

The GVRP-DWH presented in this work is capable of considering a variety of real-life requirements, in particular, certain regulations concerning drivers' working hours. However, further research is necessary in order to consider all of the regulations imposed by EU social legislation. Real-life problems may encounter complexities not considered by the GVRP-DWH and future research will be required in order to consider additional constraints. The GVRP-DWH provides a good starting point for such extensions.

The Reduced Variable Neighbourhood Search algorithm can be improved by adding new neighbourhood structures. The RVNS algorithm randomly chooses the next neighbourhood structure to be used. A more effective way of determining when to use which neighbourhood remains for future research. The Large Neighbourhood Search algorithms presented randomly choose the number of transportation requests to be removed. Future research should give insight in how the number of removals can be chosen more effectively. The relatedness measure proposed works well for the GVRP, however, it is counter productive for the GVRP-DWH. Future research is necessary to develop effective and efficient relatedness measures for the GVRP-DWH. LNS appears to be particularly suited for parallelisation, future research should investigate the respective potential. Extensive field studies should be conducted in order to measure the performance of the proposed methods in real-life applications.

The methods presented in this work make use of conditions (C1) and (C2) and the resulting fact that every removal from a feasible tour is feasible. Future research, should develop methods for handling problem instances where these properties are not satisfied. This would allow to tackle problems with multiple resource layers, e.g. in which a trailer may need to be assigned to a truck before assigning shipments to the vehicle.

References

E.H.L. Aarts and J.K. Lenstra. *Local Search in Combinatorial Optimization.* Wiley, Chichester, United Kingdom, 1997.

R.K. Ahuja, O. Ergun, J.B. Orlin, and A.P. Punnen. A survey of very large-scale neighborhood search techniques. *Discrete Applied Mathematics*, 123:75–102, 2002.

R.K. Ahuja, T.L. Magnanti, and J.B. Orlin. *Network Flows: Theory, Algorithms and Applications*. Prentice Hall, New York, 1993.

Airports Council International. Cargo Traffic: Januar 2004 - September 2004. http://www.airports.org [last accessed 04.01.2005], 2005.

J. Antes and U. Derigs. A new parallel tour construction algorithm for the vehicle routing problem with time windows. Department of Information Systems and Operations Research, University of Cologne, Cologne, Germany, 1995.

Autobahn NRW. World Wide Web. http://www.autobahn.nrw.de [last accessed 16.11.2005], 2005.

S. Baumann. Einsatz von Navigations-, Kommunikations- und Erdbeobachtungssatelliten in der Verkehrstelematik. Dissertation, Mathematisch-Naturwissenschaftliche Fakultät, Heinrich-Heine-Universität Düsseldorf, 2002.

S. Becker. Reorganisation der Luftfrachtkonsolidierung: Eine Analyse strategischer Allianzen von Luftfrachtspeditionen. Vandenhoeck & Rupprecht, Göttingen, 1999.

W. Benning and T. Aussems. Mobile mapping by a car driven survey system (CDSS). In H. Kahmen, editor, *Symposium on Geodesy for Geotechnical and Structural Engineering, Eisenstadt (A)*. Geodätisches Institut der RWTH Aachen, Aachen, 1998.

D. Bernstein and A. Kornhauser. An introduction to map matching for personal navigation assistants. Technical report, New Jersey TIDE Center, New Jersey Institute of Technology, Newark, New Jersey, 1996.

C. Bierwirth, S. Schneider, and H. Kopfer. Elektronische Transportmärkte - Aufgaben, Entwicklungsstand und Gestaltungsoptionen. *Wirtschaftsinformatik*, 44: 335–344, 2002.

C. Blum and A. Roli. Metaheuristics in combinatorial optimization: Overview and conceptual comparison. *ACM Computing Surveys*, 35(3):268–308, 2003.

L.D. Bodin. Twenty years of routing and scheduling. *Operations Research*, 38(4): 571–579, 1990.

A.J. Bower. Audio in digital radio - the Eureka 147 DAB system. *Electronic Engineering*, pages 55–56, 1998.

Britannica Online. World Wide Web. http://www.britannica.com [last accessed 08.11.2005], 2005.

British Airways World Cargo. Worldwide Timetable: November 2004 - March 2005. http://www.baworldcargo.com [last accessed 05.01.2005], 2004.

O. Bräysy and M. Gendreau. Vehicle routing problem with time windows, Part I: Route construction and local search algorithms. *Transportation Science*, 39(1): 104–118, 2005a.

O. Bräysy and M. Gendreau. Vehicle routing problem with time windows, Part II: Metaheuristics. *Transportation Science*, 39(1):119–139, 2005b.

A. Campbell and M. Savelsbergh. Efficient insertion heuristics for vehicle routing and scheduling problems. *Transportation Science*, 38(3):369–378, 2004.

Z.-L. Chen and H. Xu. Dynamic column generation for dynamic vehicle routing with time windows. *Transportation Science*, 40(1):74–88, 2006.

G. Clarke and J.V. Wright. Scheduling of vehicles from a central depot to a number of delivery points. *Operations Research*, 12:568–581, 1964.

W. Cohen. Taking to the highway - the trucking industry transforms itself with cutting-edge technology. US News and World Report 18.09.1995, 1995.

Comité Européen de Normalisation: Technical Committee 278. GDF 3.0 Manual. http://www.ertico.com/en/links/links/gdf_-_geographic_data_files.htm [last accessed 08.11.2005], 1995.

G. Comparetto and R. Ramirez. Trends in mobile satellite technology. *IEEE Computer*, 30(2):44–52, 1997.

J.-F. Cordeau, M. Gendreau, A. Hertz, G. Laporte, and J.-S. Sormany. New heuristics for the vehicle routing problem. Les cahiers du GERAD G-2004-33, Université de Montréal HEC, Montréal, Canada, 2004.

J.-F. Cordeau and G. Laporte. The dial-a-ride problem (DARP): Variants, modeling issues and algorithms. *4OR: Quarterly Journal of the Belgian, French and Italian Operations Research Societies*, 1(2):89–101, 2003.

Council of the European Communities. Council Regulation (EEC) No 3820/85 of 20 December 1985 on the harmonization of certain social legislation relating to road transport, 1985.

Council of the European Communities. Council Regulation (EEC) No 3118/93 of 25 October 1993 laying down the conditions under which non-resident carriers may operate national road haulage services within a Member State, 1993.

Council of the European Union. Council Regulation (EC) No 3315/94 of 22 December 1994 amending Regulation (EEC) No 3118/93 of 25 October 1993 laying down the conditions under which non-resident carriers may operate national road haulage services within a Member State, 1994.

R. Czommer. Leistungsfähigkeit fahrzeugautonomer Ortungsverfahren auf der Basis von Map-Matching-Techniken. Dissertation, Institut für Anwendungen der Geodäsie im Bauwesen, Universität Stuttgart, 2000.

G.B. Dantzig and J.H. Ramser. The truck dispatching problem. *Management Science*, 6:80–91, 1959.

datafactory AG. DATAfleet ODBC Interface: Programmers Guide. DATAfleet Documentation, 2002.

datafactory AG. World Wide Web. http://www.datafactory.de [last accessed 28.03.2005], 2005.

Department of the Environment. Handling geographic information. Report of the Committee of Enquiry chaired by Lord Chorley. HMSO, London, 1987.

G. Desaulniers, J. Desrosiers, A. Erdmann, M.M. Solomon, and F. Soumis. VRP with pickup and delivery. In P. Toth and D. Vigo, editors, *The Vehicle Routing Problem*, pages 225–242. SIAM Monographs on Discrete Mathematics and Applications, Philadelphia, 2002.

M. Desrochers, J. Desrosiers, and M. Solomon. New optimization algorithm for the vehicle routing problem with time windows. *Operations Research*, 40(2):342–354, 1992.

DEUFRAKO. IVHW System Concept and Issues relevant for Standardization. DEUFRAKO Project Consortium - Draft, 2002.

C. Drane, M. Macnaughtan, and C. Scott. Positioning GSM telephones. *IEEE Communications Magazine*, pages 46–59, 1998.

Y. Dumas, J. Desrosiers, and F. Soumis. The pickup and delivery problem with time windows. *European Journal of Operational Research*, 54:7–22, 1991.

Electronic Communications Committee. ECC Decision of 15 March 2002 on the frequency bands to be designated for the co-ordinated introduction of Road Transport and Traffic Telematic Systems, ECC/DEC/(02)01, 2002.

R. Elmasri and S.B. Navathe. Fundamentals of database systems. Third Edition, Addison-Wesley, Reading, Massachusetts, 2000.

European Commission. White paper - European transport policy for 2010: time to decide. Office for Official Publications of the European Communities, 2001.

European Commission. Europe at a crossroads - The need for sustainable transport. Office for Official Publications of the European Communities, 2003a.

European Commission. The Treaty of Accession to the European Union 2003. http://europa.eu.int/comm/enlargement/negotiations/treaty_of_accession_2003/index.htm [last accessed 29.08.2005], 2003b.

European Commission: Directorate General for Energy And Transport. European energy and transport - Trends to 2030. Office for Official Publications of the European Communities, 2003.

European Commission: Eurostat. Panorama of transport - Statistical overview of transport in the European Union. Office for Official Publications of the European Communities, 2003.

European Radiocommunications Committee. ERC Decision of 24th October 1994 on the frequency bands to be designated for the coordinated introduction of the GSM digital pan-European communications system, ERC/DEC/(94)01, 1994.

European Radiocommunications Committee. ERC Decision of 7 March 1996 on the frequency bands for the introduction of the Trans European Trunked Radio System (TETRA), ERC/DEC/(96)04, 1996.

European Radiocommunications Committee. ERC Decision of 30 June 1997 on the frequency bands for the introduction of the Universal Mobile Telecommunications System (UMTS), ERC/DEC/(97)07, 1997.

European Union. Regulation (EC) No 561/2006 of the European Parliament and of the Council of 15 March 2006 on the harmonisation of certain social legislation relating to road transport and amending Council Regulations (EEC) No 3821/85 and (EC) No 2135/98 and repealing Council Regulation (EEC) No 3820/85. Official Journal of the European Union L 102, 11.04.2006, 2006.

D. Feillet, P. Dejax, and M. Gendreau. Traveling Salesman Problems with Profits. *Transportation Science*, 39(2):188–205, 2005.

B. Fleischmann, S. Gnutzmann, and E. Sandvoß. Dynamic vehicle routing based on on-line traffic information. *Transportation Science*, 38(4):420–433, 2004.

FMS-Standard Working Group. FMS-Standard Interface description Version 01.00. http://www.fms-standard.com [last accessed 16.11.2005], 2002.

R.L. Freeman. *Radio system design for telecommunications (1-100 GHz)*. John Wiley & Sons, New York, 1987.

Frost & Sullivan. Analysis of the European commercial vehicle telematics market. Frost & Sullivan Report B042, 2002.

M. Gendreau, F. Guertin, J.-Y. Potvin, and R. Séguin. Neighborhood search heuristics for a dynamic vehicle dispatching problem with pick-ups und deliveries. Technical report CRT-98-10, Centre de recherche sur les transports, Université de Montréal, Montréal, 1998.

M. Gendreau, F. Guertin, J.-Y. Potvin, and É.D. Taillard. Parallel tabu search for real-time vehicle routing and dispatching. *Transportation Science*, 33(4):381–390, 1999.

M. Gendreau, G. Laporte, and J.-Y. Potvin. Metaheuristics for the capacitated VRP. In P. Toth and D. Vigo, editors, *The Vehicle Routing Problem*, pages 129–154. SIAM Monographs on Discrete Mathematics and Applications, Philadelphia, 2002.

D. Gillette. Combining communications and computing: Telematics infrastructures. In J.H. Ausubel and R. Herman, editors, *Cities and their vital systems: Infrastructure past, present, and future*, pages 233–248. National Academy Press, Washington, 1988.

A. Goel and V. Gruhn. Integration of telematics for efficient management of carrier operations. In *Proceedings of the IEEE International Conference on e-Business Engineering (ICEBE 2005)*, pages 404–408, 2005. ISBN 0-7695-2430-3.

A. Goel and V. Gruhn. Drivers' working hours in vehicle routing and scheduling. In *Proceedings of the 9th International IEEE Conference on Intelligent Transportation Systems (ITSC 2006)*, pages 1280–1285, 2006a. ISBN 1-4244-0094-5.

A. Goel and V. Gruhn. A fleet monitoring system for advanced tracking of commercial vehicles. In *Proceedings of the 2006 IEEE International Conference on*

Systems, Man, and Cybernetics (SMC 2006), pages 2517–2522, 2006b. ISBN 1-4244-0100-3.

A. Goel and V. Gruhn. An operations support system for truckage companies. In *Proceedings of the International Conference on Information Systems, Logistics, and Supply Chain (ILS 2006)*, pages 1213–1222, 2006c. ISBN 2-930294-17-5.

A. Goel and V. Gruhn. A General Vehicle Routing Problem. European Journal of Operational Research (in press), http://dx.doi.org/10.1016/j.ejor.2006.12.065, 2007.

P. Hansen and N. Mladenović. A tutorial on Variable Neighborhood Search. Les cahiers du GERAD G-2003-46, Université de Montréal HEC, Montréal, Canada, 2003.

S. Harrison. Scheduling for milk collection. Presented at the Workshop on Mathematical Models for Optimizing Transportation Services, 19.-22.04.2005, University of Auckland, Auckland, New Zealand, 2005.

H. Hartenstein, H. Füßler, M. Mauve, and W. Franz. Simulation results and a proof-of-concept implementation of the FleetNet position-based router. In *Personal Wireless Communications*, volume 2775 of *Lecture Notes in Computer Science*, pages 192–197. Springer-Verlag, 2003.

M. Heckmann. DV-gestütztes Geschäftsprozeßmanagement in der Luftfrachtlogistik. Dissertation, Shaker Verlag Aachen, 2002.

T. Herrmann, L. Schöpe, E. Erkens, and M. Hülder. Mobile Speditionslogistikunterstützung. Schlussbericht des Verbundforschungsprojekt SpiW "Mobile Spedition im Web", Shaker Verlag Aachen, 2005.

S. Ichoua, M. Gendreau, and J.-Y. Potvin. Diversion issues in real-time vehicle dispatching. *Transportation Science*, 34(4):426–438, 2000.

ILOG. ILOG CPLEX 7.1 User's manual, 2001.

Inmarsat. World Wide Web. http://www.inmarsat.org [last accessed 08.11.2005], 2005.

International Organisation for Standardization. Transport Information and Control Systems - Reference Model Architecture(s) for the TICS sector - Part 1: Fundamental TICS services. ISO/TC204/WG1/N310R, 1997.

International Organisation for Standardization. Traffic and Traveller Information (TTI) - TTI messages via traffic message coding - Part 1: Coding protocol for Radio Data System - Traffic Message Channel (RDS-TMC) using ALERT-C. ISO 14819-1:2003, 2003.

International Organisation for Standardization. Intelligent transport systems - Geographic Data Files (GDF) - Overall data specification. ISO 14825:2004, 2004.

M. E. Johnson and D. F. Pyke. Supply chain management: Integration and globalization in the age of ebusiness. Tuck school of business working paper no. 02-09, Tuck School of Business at Dartmouth, Hanover, New Hampshire, 2001.

P. Kilby, P. Prosser, and P. Shaw. Dynamic VRPs: A study of scenarios. APES report APES-06-1998, APES Research Group, 1998.

P. Kilby, P. Prosser, and P. Shaw. A comparison of traditional and constraint-based heuristic methods on vehicle routing problems with side constraints. *Constraints*, 5:389–414, 2000.

S. Kirkpatrick, C.D. Gelatt, and M.P. Vecchi. Optimization by simulated annealing. *Science*, 220(4598):671–680, 1983.

H. Kopfer and J. Schönberger. Interactive solving of vehicle routing and scheduling problems: Basic concepts and qualification of tabu search approaches. In *HICSS '02: Proceedings of the 35th Annual Hawaii International Conference on System Sciences (HICSS-35)-Volume 3*, page 84. IEEE Computer Society, 2002.

D. Kopitz and B. Marks. Traffic and Travel Information broadcasting - protocols for the 21st century. EBU Technical Review, http://www.ebu.ch/en/technical/trev/trev_279-contents.html [last accessed 27.12.2005], 1999.

F. Lafontaine and L. Malaguzzi Valeri. The deregulation of international trucking in the European Union: Form and effect. Michigan law and economics working paper no. 05-006, John M. Olin Center for Law & Economics, University of Michigan, 2005.

G. Laporte and F. Semet. Classical heuristics for the capacitated VRP. In P. Toth and D. Vigo, editors, *The Vehicle Routing Problem*, pages 109–128. SIAM Monographs on Discrete Mathematics and Applications, Philadelphia, 2002.

A. Larsson. The dynamic vehicle routing problem. Ph.D. thesis, Technical University of Denmark (IMM), Lyngby, Denmark, 2000.

R. Lasch. Telematik in der Güterverkehrslogistik - Stand und Perspektiven. In: Kundenbindung bei veränderten Wettbewerbsbedingungen, Teubner Stuttgart, 2000.

E.L. Lawler, J.K. Lenstra, A.H.G. Rinnooy Kan, and D.B. Shmoys. *The Traveling Salesman Problem.* Wiley, Chichester, United Kingdom, 1985.

H.L. Lee and S. Whang. E-business and supply chain integration. White paper, SGSCMF- W2-2001, Stanford Global Supply Chain Management Forum, Stanford University, Stanford, California, 2001.

S.M. Lewandowski. Frameworks for component-based client/server computing. *ACM Computing Surveys*, 30(1):3–27, 1998.

H. Li and A. Lim. A metaheuristic for the pickup and delivery problem with time windows. In *Proceedings of the 13th IEEE International Conference on Tools with Artificial Intelligence (ICTAI'01)*, pages 160–167, 2001.

S. Lin. Computer solutions of the traveling salesman problem. *Bell Systems Technology Journal*, 44:2245–2269, 1965.

P.A. Longley, M.F. Goodchild, D.J. Maguire, and Rhind D.W. *Geographic Information Systems and Science*. John Wiley & Sons, West Sussex, 2001.

H.R. Lourenço, O.C. Martin, and T. Stützle. Iterated Local Search. In F. Glover and G. Kochenberger, editors, *Handbook of Metaheuristics*, pages 321–353. Kluwer, 2002.

Q. Lu and M. M. Dessouky. A new insertion-based construction heuristic for solving the pickup and delivery problem with time windows. *European Journal of Operational Research*, 175:672–687, 2006.

M.L. Manheim. Opportunities for competitive advantage in logistics through information technology. In *Proceedings of the 29th Annual Hawaii International Conference on System Sciences*, pages 440–449, 1996.

A.J. Mason. Emergency vehicle trip analysis using GPS AVL data: A dynamic program for map matching. In *Proceedings of the 40th Annual Conference of the*

Operational Research Society of New Zealand, 2-3 December 2005, Wellington, New Zealand, pages 295–304, 2005.

K. Menger. Das Botenproblem. In K. Menger, editor, *Ergebnisse eines Mathematischen Kolloquiums 2*, pages 11–12. Teubner, Leipzig, Germany, 1932.

H.J. Miller and S.-L. Shaw. *Geographic Information Systems for Transportation*. Oxford University Press, New York, 2001.

S. Mitrović-Minić. Pickup and delivery problem with time windows: A survey. Technical report TR 1998-12, School of Computing Science, Simon Fraser University, Burnaby, BC, Canada, 1998.

S. Mitrović-Minić. The dynamic pickup and delivery problem with time windows. Ph.D. thesis, School of Computing Science, Simon Fraser University, Burnaby, BC, Canada, 2001.

N. Mladenović and P. Hansen. Variable Neighborhood Search. *Computers and Operations Research*, 24:1097–1100, 1997.

R. Montemanni, L.M. Gambardella, A.E. Rizzoli, and A.V. Donati. A new algorithm for a dynamic vehicle routing problem based on ant colony system. Tech report IDSIA-23-02, Istituto Dalle Molle di Studi sull'Intelligenza Artificiale (IDSIA), Lugano, Switzerland, 2002.

S. Nora and A. Minc. *L'Informatisation de la société. Rapport à M. le Président de la République*. La Documentation Française, Paris, 1978.
English translation: *The Computerization of Society. A report to the President of France*. MIT Press, Cambridge, Massachusetts, 1980.

I. Or. Travelling salesman-type combinatorial problems and their relation to the logistics of blood-banking. Ph.D. thesis, Department of Industrial Engineering and Management Sciences, Northwest University, Evanston, Illinois, 1976.

Oracle. World Wide Web. http://www.oracle.com [last accessed 19.06.2006], 2006.

ORBCOMM. World Wide Web. http://www.orbcomm.com [last accessed 08.11.2005], 2005.

Organisation for economic co-operation and development. Liberalisation and structural reform in the freight transport sector in Europe. http://www1.oecd.org/ech/pub/TRANSp1.PDF [last accessed 29.08.2005], 1997.

G. Pankratz. Dynamic planning of pickup and delivery operations by means of genetic algorithms. Diskussionsbeitrag Nr. 361, Fachbereich Wirtschaftswissenschaften, Fernuniversität Hagen, Hagen, Germany, 2004.

PIARC Commitee on Intelligent Transport. *ITS Handbook 2000 - Recommendations from the World Road Association (PIARC)*. Artech House, London, 1999.

M. Polacek, R.F. Hartl, K. Doerner, and M. Reimann. A variable neighborhood search for the multi depot vehicle routing problem with time windows. *Journal of Heuristics*, 10:613–627, 2004.

J.-Y. Potvin and J.-M. Rousseau. A parallel route building algorithm for the vehicle routing and scheduling problem with time windows. *European Journal of Operations Research*, 66:331–340, 1993.

W.B. Powell. Dynamic models of transportation operations. In S. Graves and T. A. G. de Tok, editors, *Handbooks in Operations Research and Management Science: Supply Chain Management*, pages 677–756. Elsevier, Amsterdam, 2003.

W.B. Powell, A. Marar, J. Gelfand, and S. Bowers. Implementing real-time optimization models: A case application from the motor carrier industry. *Operations Research*, 50(4):571–581, 2002.

W.B. Powell, M.T. Towns, and A. Marar. On the value of optimal myopic solutions for dynamic routing and scheduling problems in the presence of user noncompliance. *Transportation Science*, 34(1):67–85, 2000.

Prognos AG. Wirkungspotentiale der Verkehrstelematik zur Verbesserung der Verkehrsinfrastruktur und Verkehrsmittelnutzung. Forschungsbericht FE-Nr. 96.584/1999, 2001.

H.N. Psaraftis. Dynamic vehicle routing problems. In B.L. Golden and A.A. Assad, editors, *Vehicle routing: Methods and studies*, pages 233–248. North-Holland Amsterdam, 1988.

H.N. Psaraftis. Dynamic vehicle routing: Status and prospects. *Annals of Operations Research*, 61:143–164, 1995.

Qualcomm. World Wide Web. http://www.qualcomm-europe.com [last accessed 08.11.2005], 2005.

M.A. Quddus, W.Y. Ochieng, L. Zhao, and R.B. Noland. A general map matching algorithm for transport telematics applications. *GPS Solutions Journal*, 7(3):157–167, 2003.

A.C. Regan, H. Mahmassani, and P. Jaillet. Real-time information for improved efficiency of commercial vehicle operations. Technical report SWUTC/98/60031-1, University of Texas at Austin, 1998.

D. Reichardt, M. Miglietta, L. Moretti, P. Morsink, and W. Schulz. CarTALK 2000: safe and comfortable driving based upon inter-vehicle-communication. In *IEEE Intelligent Vehicle Symposium*, pages 545–550, 2002.

S. Ropke and D. Pisinger. An adaptive large neighborhood search heuristic for the pickup and delivery problem with time windows. *Transportation Science*, 40(4): 455–472, 2006.

J. Roy. Recent trends in logistics and the need for real-time decision tools in the trucking industry. In *HICSS '01: Proceedings of the 34th Annual Hawaii International Conference on System Sciences (HICSS-34)-Volume 3*, page 3031. IEEE Computer Society, 2001.

M.W.P. Savelsbergh. The vehicle routing problem with time windows: Minimizing route duration. *ORSA Journal on Computing*, 4:146–154, 1992.

M.W.P. Savelsbergh and M. Sol. The general pickup and delivery problem. *Transportation Science*, 29(1):17–30, 1995.

M.W.P. Savelsbergh and M. Sol. DRIVE: dynamic routing of independent vehicles. *Operations Research*, 46:474–490, 1998.

G. Schrimpf, J. Schneider, H. Stamm-Wilbrandt, and G. Dueck. Record breaking optimization results using the ruin and recreate principle. *Journal of Computational Physics*, 159:139–171, 2000.

J. Schönberger, H. Kopfer, and H. Mattfeld. A combined approach to solve the pickup and delivery selection problem. In *Operations Research Proceedings 2002*, pages 150–155. Springer, 2002.

P. Shaw. A new local search algorithm providing high quality solutions to vehicle routing problems. Technical report, APES group, Department of Computer Sciences, University of Strathclyde, Glasgow, Scottland, 1997.

Y. Shen, J.-Y. Potvin, J.-M. Rousseau, and S. Roy. A computer assistant for vehicle dispatching with learning capabilities. *Annals of Operations Research*, 61:189–211, 1995.

M.M. Solomon. Algorithms for the vehicle routing and scheduling problem with time window constraints. *Operations Research*, 35:254–265, 1987.

M.Z. Spivey and W.B. Powell. The dynamic assignment problem. *Transportation Science*, 38(4):399–419, 2004.

K. Srinavasan, S. Kekre, and T. Mukhopadhyay. Impact of electronic data interchange technology on JIT shipments. *Management Science*, 10:1291–1304, 1994.

É.D. Taillard. A heuristic column generation method for the heterogeneous fleet VRP. Technical report CRT-96-03, Centre de recherche sur les transports, Université de Montréal, Montréal, 1996.

É.D. Taillard, L.M. Gambardella, M. Gendreau, and J.-Y. Potvin. Adaptive memory programming: A unified view of metaheuristics. Tech report IDSIA-19-98, Istituto Dalle Molle di Studi sull'Intelligenza Artificiale (IDSIA), Lugano, Switzerland, 1998.

G. Thornton. Optimistic locking with concurrency in oracle. http://www.orafaq.org/papers/locking.pdf [last accessed 03.09.2005], 2001.

TollCollect. World Wide Web. http://www.tollcollect.de [last accessed 16.11.2005], 2005.

P. Toth and D. Vigo. *The Vehicle Routing Problem.* SIAM Monographs on Discrete Mathematics and Applications, Philadelphia, 2002.

S. Tsugawa. Issues and recent trends in vehicle safety communication systems. *IATTS Research*, 29(1):7–15, 2005.

E. Turban, D. Leidner, E. McLean, and J. Wetherbe. *Information Technology for Management: Transforming Organizations in the Digital Economy.* John Wiley & Sons, Inc., 5th edition, 2005.

United Nations Conference on Trade and Development. UNCTAD Handbook of Statistics 2003, 2003.

United Nations Conference on Trade and Development. UNCTAD Development and Globalization: Facts and Figures, 2004a.

United Nations Conference on Trade and Development. UNCTAD Trade and Development Report, 2004b.

K.-J. van Dorp. Tracking and tracing: a structure for development and contemporary practices. *Journal of Enterprise Information Management*, 15(1):24–33, 2002.

Verein Bremer Spediteure e.V. Jahresbericht 2005 über die Lage des Speditionsgewerbes. http://www.vbsp.de/jahresbericht_2005.pdf [last accessed 10.09.2005], 2005.

M. Wardman, A. Bristow, J. Toner, and G. Tweddle. Review of research relevant to rail competition for short haul air routes. Eurocontrol, Report SEE/2002/003, 2002.

C.E. White, D. Bernstein, and A.L. Kornhauser. Some map matching algorithms for personal navigation assistants. *Transportation Research Part C*, 8:91–108, 2000.

N.H.M. Wilson, J.M. Sussman, H.-K. Wong, and T. Higonnet. Scheduling algorithms for a dial-a-ride-system. Urban Systems Laboratory Report USL TR-70-13, Massachusetts Institute of Technology (MIT), 1970.

L. Wood, P. Worfolk, and R. Thurman. SaVi - software for satellite visualization. The Geometry Center, University of Minnesota, 1996.

H. Xu, Z.-L. Chen, S. Rajagopal, and S. Arunapuram. Solving a practical pickup and delivery problem. *Transportation Science*, 37(3):347–364, 2003.

Y. Xu. Development of transport telematics in Europe. *GeoInformatica*, 4(2):179–200, 2000.

J. Yang, P. Jaillet, and H. Mahmassani. On-line algorithms for the truck fleet assignment and scheduling under real-time information. *Transportation Research Record 1667*, pages 107–113, 1999.

J. Yang, P. Jaillet, and H. Mahmassani. Real-time multi-vehicle truckload pickup-and-delivery problems. *Transportation Science*, 38(2):135–148, 2004.

Abbreviations

AMP	Adaptive Memory Programming
BS	Billing System
CPAS	Cost & Performance Analysis System
DAB	Digital Audio Broadcasting
DPS	Dynamic Planning System
DSRC	Dedicated Short Range Communication
EFM	Electronic Freight Market
ETC	Electronic Toll Collection
EU	European Union
EU-15	European Union including the 15 member states before the enlargement in 2004
EU-25	European Union including the 25 member states after the enlargement in 2004
FTS	Fleet Telematics System
FCS	Fleet Communication System
FTL	Full-Truckload
GDF	Geographic Data File
GDP	Gross Domestic Product
GEO	Geostationary orbit
GIS	Geographical Information System
GIS-T	Geographical Information System for Transportation

GLONASS	Global'naya Navigatsionnaya Sputnikovaya Sistema
GNSS	Global Navigation Satellite System
GPS	Global Positioning System
GSM	Global System for Mobile Communication
GUI	Graphical User Interface
GVRP	General Vehicle Routing Problem
GVRP-DWH	General Vehicle Routing Problem with Drivers' Working Hours
HFVRP	Heterogeneous Fleet Vehicle Routing Problem
HFVRPTW	Heterogeneous Fleet Vehicle Routing Problem with Time Windows
ILS	Iterated Local Search
ITS	Intelligent Transportation Systems
LAFES	Load Acquisition & Freight Exchange System
LEO	Low-Earth orbit
LIS	Legacy Information System
LNS	Large Neighbourhood Search
LTL	Less-Than-Truckload
MFMS	Messaging & Fleet Monitoring System
MIS	Management Information System
OFMS	Order & Fleet Management System
Q.E.D.	Quod Erat Demonstrandum
PDP	Pickup and Delivery Problem
PDPTW	Pickup and Delivery Problem with Time Windows
PIN	Personal Identification Number
RDS	Radio Data System
RFS	Road Feeder Services
RVNS	Reduced Variable Neighbourhood Search
SCM	Supply Chain Management
SPS	Static Planning System
TICS	Transport Information and Control Systems

TCN	Transaction Control Number
TMC	Traffic Message Channel
TSP	Travelling Salesman Problem
TTIS	Traffic & Travel Information Systems
ULD	Unit Load Device
UMTS	Universal Mobile Telecommunication System
VNS	Variable Neighbourhood Search
VRP	Vehicle Routing Problem
VRPTW	Vehicle Routing Problem with Time Windows
VRP-DWH	Vehicle Routing Problem with Drivers' Working Hours
VS	Vehicle System

Symbols

$\mathcal{A}$	Set of arcs in a network
$\mathcal{A}_\theta$	Set of arcs used by tour θ
(A)	Assumption that for all vehicles $v \in \mathcal{V}$: $\theta_v := (n_{(v,1)}, \ldots, n_{(v,\lambda_v)})$ is a feasible tour (see page 129)
$\mathcal{C}$	Set of nodes corresponding to customer locations
(C1)	Assumption that $\sum_{\mu=1}^{\mu \leq i} r_{n_{(o,\mu)}} \geq 0$ for all orders $o \in \mathcal{O}$ and all $1 \leq i \leq \lambda_o$ (see page 134)
(C2)	Assumption that $\sum_{\mu=1}^{\mu \leq i} r_{n_{(o,\mu)}} \leq 0$ for all orders $o \in \mathcal{O}$ and all $1 \leq i \leq \lambda_o$ (see page 134)
c_{nm}	Cost for travelling from $n \in \mathcal{N}$ to $m \in \mathcal{N}$
c_{nm}^v	Cost of vehicle $v \in \mathcal{V}$ for travelling from node $n \in \mathcal{N}$ to $m \in \mathcal{N}$
c_θ	Cost of tour θ
$\mathcal{D}$	Set of nodes usually corresponding to the depot(s) or the vehicles' current locations
δ_{ov}	Binary parameter indicating whether $o \in \mathcal{O}$ can be served by $v \in \mathcal{V}$ ($\delta_{ov} = 1$), or not ($\delta_{ov} = 0$)
δ_{nm}^v	Pure driving time of vehicle $v \in \mathcal{V}$ from node $n \in \mathcal{N}$ to $m \in \mathcal{N}$
d_{nm}	Travel time (duration) from node $n \in \mathcal{N}$ to $m \in \mathcal{N}$
d_{nm}^v	Travel time (duration) of vehicle $v \in \mathcal{V}$ from node $n \in \mathcal{N}$ to $m \in \mathcal{N}$
λ_o	Number of nodes belonging to $o \in \mathcal{O}$
λ_v	Number of nodes vehicle $v \in \mathcal{V}$ must visit
$\mathscr{L}$	Set of labels

$\mathscr{L}_m(l_n)$	Set of feasible labels at node $m \in \mathcal{N}$, for a vehicle coming from node $n \in \mathcal{N}$ with label l_n
l_n	Label indicating the driver state at node $n \in \mathcal{N}$
$\mathcal{N}$	Set of nodes in a network
$N_i(s)$	The ith neighbourhood of solution s
n, m	Node in a network $\mathcal{N}$
n_{depot}	Node corresponding to the depot
n_v	Node corresponding to the depot of vehicle $v \in \mathcal{V}$
$n_{(v,i)}$	Node corresponding to the ith location vehicle $v \in \mathcal{V}$ must visit
$n_{(o,i)}$	Node corresponding to the ith location belonging to order $o \in \mathcal{O}$
$\mathcal{O}$	Set of orders
$\mathcal{O}_\theta$	Set of orders served by tour θ
o	An order
p_o	The revenue of order $o \in \mathcal{O}$
ρ_n	The accumulated load at node $n \in \mathcal{N}$
r_n	The resource demand (or supply) at at node $n \in \mathcal{N}$
$r^{\max}$	The capacity of all vehicles
r^v	The capacity of vehicle $v \in \mathcal{V}$
s, s', s^*	A solution of the optimisation problem
s_n^v	The service time required at node $n \in \mathcal{N}$ for vehicle $v \in \mathcal{V}$
θ	Sequence of nodes representing a tour
t_n	Arrival time at node $n \in \mathcal{N}$
$t_n^{\min}$	Lower bound on the arrival time at node $n \in \mathcal{N}$
$t_n^{\max}$	Upper bound on the arrival time at node $n \in \mathcal{N}$
t^v_{break}	The time required for a break for vehicle $v \in \mathcal{V}$
t^v_{daily}	The maximum daily driving time between two consecutive daily rest periods for vehicle $v \in \mathcal{V}$
t^v_{nonstop}	The maximum nonstop driving time between two consecutive breaks or rest periods for vehicle $v \in \mathcal{V}$
t^v_{rest}	The time required for a daily rest period for vehicle $v \in \mathcal{V}$

t^v_{weekly}	The maximum weekly driving time between two consecutive weekly rest periods for vehicle $v \in \mathcal{V}$
$\mathcal{V}$	Set of vehicles
v	A vehicle
x_{nm}	Binary variable indicating whether a vehicle travels from node $n \in \mathcal{N}$ to node $m \in \mathcal{N}$ ($x_{nm} = 1$), or not ($x_{nm} = 0$)
x^v_{nm}	Binary variable indicating whether vehicle $v \in \mathcal{V}$ travels from node $n \in \mathcal{N}$ to node $m \in \mathcal{N}$ ($x^v_{nm} = 1$), or not ($x^v_{nm} = 0$)
y^v_n	Binary variable indicating whether vehicle $v \in \mathcal{V}$ visits node $n \in \mathcal{N}$ ($y^v_n = 1$), or not ($y^v_n = 0$)

Index

www.ingramcontent.com/pod-product-compliance
Lightning Source LLC
Chambersburg PA
CBHW072203090125
20170CB00003B/81
* 9 7 8 0 3 8 7 7 5 1 0 4 7 *